Matrix Theory

Matrix Theory

DAVID W. LEWIS
Department of Mathematics
University College Dublin
Ireland

Published by

World Scientific Publishing Co. Pte. Ltd.
P O Box 128, Farrer Road, Singapore 9128
USA office: Suite 1B, 1060 Main St., River Edge, NJ 07661
UK office: 73 Lynton Mead, Totteridge, London N20 8DH

Library of Congress Cataloging-in-Publication data is available.

MATRIX THEORY

ISBN 981-02-0689-5

Printed in Singapore by Utopia Press.

PREFACE

This book provides an introduction to matrix theory, and aims to provide a clear and concise exposition of the basic ideas, results and techniques in the subject. It combines the algebraic and analytic aspects of matrix theory. It presumes no knowledge beyond school mathematics although some familiarity with elementary calculus would be helpful in a few of the applications. It is hoped that the book can be profitably used by a wide range of students, including students of mathematics, engineering, science, and other disciplines where matrices arise. Complete proofs are given, although some are relegated to appendices at the end of chapters. This should enable the book to be used both by students who want all of the theory and those who are mainly interested in learning the techniques. The text is interspersed with many examples, applications and numerous exercises for the reader. Students who have already done an introductory linear algebra course may use the later chapters for a more advanced course.

Acknowledgements

I am indebted to my colleagues Fergus Gaines, Rod Gow and Tom Laffey for many useful discussions on matrices and linear algebra, and for reading the text and correcting some errors. I am grateful to my wife Anne for her constant support and encouragement during the preparation of this book.

TABLE OF CONTENTS

Chapter 7 Perturbation theory [illegible]

Chapter 8 Further topics

Chapter 1

MATRICES AND LINEAR EQUATIONS

A familiarity with matrices is necessary nowadays in many areas of mathematics and in a wide variety of other disciplines. Areas of mathematics where matrices occur include algebra, differential equations, calculus of several variables, probability and statistics, optimization, and graph theory. Other disciplines using matrix theory include engineering, physical sciences, biological sciences, economics and management science.

In this first chapter we give the fundamentals of matrix algebra, determinants, and systems of linear equations. At the end of the chapter we give some examples of situations in mathematics and other disciplines where matrices arise.

1.1 Matrices and matrix algebra

A *matrix* is a rectangular array of symbols. In this book the symbols will usually be either real or complex numbers. The separate elements of the array are known as the *entries* of the matrix.

Let m and n be positive integers. An m×n matrix A consists of m rows and n columns of numbers written in the following manner.

$$A = \begin{pmatrix} a_{11} & a_{12} & \cdots & a_{1n} \\ a_{21} & a_{22} & \cdots & a_{2n} \\ \vdots & & & \vdots \\ a_{m1} & a_{m2} & \cdots & a_{mn} \end{pmatrix}$$

We often write $A = (a_{ij})$ for short. The entry a_{ij} lies in the i-th row and the j-th column of the matrix A.

Two m×n matrices $A = (a_{ij})$ and $B = (b_{ij})$ are *equal* if and only if all the corresponding entries of A and B are equal.

i.e. $a_{ij} = b_{ij}$ for each i and j.

The *sum* of the m×n matrices $A = (a_{ij})$ and $B = (b_{ij})$ is the m×n matrix denoted A + B which has entry $a_{ij} + b_{ij}$ in the (i,j)-place for each i,j.

Let λ be a scalar (i.e. a real or complex number) and let $A = (a_{ij})$ be an m×n matrix. The *scalar multiple* of A by λ is the m×n matrix denoted λA which has entry λa_{ij} in the (i,j)-place for each i,j.

1.1.2 **Proposition**

The following properties hold.

(i) A + B = B + A for all m×n matrices A and B, i.e. addition of matrices is commutative.

(ii) (A + B) + C = A + (B + C) for all m×n matrices A,B, and C, i.e.addition of matrices is associative.

(iii) $\lambda(A + B) = \lambda A + \lambda B$ for all scalars λ and all m×n matrices A and B.

(iv) $(\lambda_1 + \lambda_2)A = \lambda_1 A + \lambda_2 A$ for all scalars λ_1,λ_2 and all m×n matrices A.

(v) $(\lambda_1\lambda_2)A = \lambda_1(\lambda_2 A)$ for all scalars λ_1,λ_2 and all mxn matrices A.

Proof

These properties follow at once from the properties of the real and complex number systems.

1.1.3 **Remark**

If we write -A for the matrix whose entries are $-a_{ij}$ for each i,j then -A = (-1)A, i.e. the multiple of the matrix A by the scalar -1. Also if we denote by O the nxn matrix with zero as each entry then A + (-A) = O.

1.1.4 Matrix multiplication

A $1\times n$ matrix will be called a *row vector of length n* and an $m\times 1$ matrix will be called a *column vector of length m.*

Let $A = (a_1\ a_2\ a_3\ .\ .\ .\ a_n)$ be a $1\times n$ matrix and $B = \begin{pmatrix} b_1 \\ b_2 \\ \vdots \\ b_n \end{pmatrix}$ be an $n\times 1$ matrix.

We define the *product* AB to be the 1×1 matrix with the single entry

$$a_1b_1 + a_2b_2 + \dots + a_nb_n.$$

Now we will define matrix multiplication in general. We say that the product AB of the two matrices A and B is defined if and only if the number of columns of A equals the number of rows of B.

(i.e. AB is defined if and only if A is an $m\times n$ matrix and B is an $n\times p$ matrix for some integers m,n,p.)

We define the *matrix product* AB to be the $m\times p$ matrix which has as its (i,j)-entry

$$a_{i1}b_{1j} + a_{i2}b_{2j} + a_{i3}b_{3j} + \dots + a_{in}b_{nj}$$

(In shorthand notation the (i,j)-entry is $\sum_{k=1}^{n} a_{ik}b_{kj}$.)

In other words the (i,j)-entry of AB is the product of the i-th row of A with the j-th column of B, this product being as in the special case of $1\times n$ and $n\times 1$ matrices defined above.

1.1.5 Example

Let $A = \begin{pmatrix} 1 & -1 & 6 \\ 4 & 1 & -2 \\ 3 & 2 & 0 \end{pmatrix}$, $B = \begin{pmatrix} -1 & 6 & 1 & 2 \\ 0 & -2 & 2 & 1 \\ 1 & 1 & 1 & 1 \end{pmatrix}$.

Then $AB = \begin{pmatrix} 1 & -1 & 6 \\ 4 & 1 & -2 \\ 3 & 2 & 0 \end{pmatrix}\begin{pmatrix} -1 & 6 & 1 & 2 \\ 0 & -2 & 2 & 1 \\ 1 & 1 & 1 & 1 \end{pmatrix} = \begin{pmatrix} 5 & 14 & 5 & 7 \\ -6 & 20 & 4 & 7 \\ -3 & 14 & 7 & 8 \end{pmatrix}$.

1.1.6 **Proposition**

The following properties hold.

(i) (AB)C = A(BC) whenever these products are meaningful.

(i.e. matrix multiplication is associative).

(ii) A(B + C) = AB + AC for all m×n matrices A,B and all n×p matrices C.

(iii) (A +B)C = AC + BC for all m×n matrices A and B and all n×p matrices C.

Proof

(i) Let A,B,C be of sizes m×n, n×p, p×q respectively.

The (i,k)-entry of AB is $\sum_{r=1}^{n} a_{ir}b_{rk}$ and hence the (i,j)-entry of (AB)C is $\sum_{r=1}^{n}\sum_{k=1}^{p} a_{ir}b_{rk}c_{kj}$. An examination of the product A(BC) shows that exactly the same expression occurs as the (i,j)-entry of A(BC).

(ii) Let A be of size m×n, B and C of size n×p.

Then the (i,j)-entry of A(B + C) is $\sum_{k=1}^{n} a_{ik}(b_{kj} + c_{kj})$ and this is easily seen to equal the (i,j)-entry of AB + AC.

(iii) This follows in a similar manner to (ii).

1.1.7 **Remark**

Matrix multiplication is not in general commutative. Note first that for AB and BA to both be defined it is necessary that A and B are each n×n matrices for some integer n, i.e. square matrices of the same size. However AB and BA will be different in general.

1.1.8 **Exercise**

Let A = $\begin{pmatrix} 1 & 2 \\ 2 & 3 \end{pmatrix}$, B = $\begin{pmatrix} 0 & 1 \\ 2 & 3 \end{pmatrix}$. Show that AB ≠ BA.

1.1.9 **Remark**

A matrix of especial importance is the n×n *identity matrix*, denoted I_n, which is defined to have entries $a_{ii} = 1$ for all i and $a_{ij} = 0$ for i ≠ j. Often when we are dealing with n×n matrices for a fixed value of n we will simply write I for the identity matrix omitting the suffix n.

For any m×n matrix A it is easy to see that $AI_n = A$ and that $I_m A = A$.

1.1.10 **The transpose of a matrix**

Let A be an m×n matrix.

The *transpose* of A is the n×m matrix with entry a_{ji} in the (i,j)-place. The transpose of A is denoted by A^t.

Note that the rows of A become the columns of A^t and vice versa.

1.1.11 **Proposition**

The transpose satisfies the following properties.

(i) $(A + B)^t = A^t + B^t$ for all m×n matrices A and B.

(ii) $(A^t)^t = A$ for all m×n matrices A.

(iii) $(AB)^t = B^t A^t$ whenever the product AB is defined.

Proof

Easy exercise.

Let A be an m×n matrix whose entries are complex numbers.

The *conjugate transpose* of A is the n×m matrix with entry $\bar{a}_{ji}$ in the (i,j)-place. The conjugate transpose is denoted $\bar{A}^t$.

The conjugate transpose satisfies the same three properties as those for the transpose given in (1.1.11).

1.1.12 **The trace of a square matrix**

Let A be an n×n matrix.

We define the *trace* of A by trace $A = \sum_{i=1}^{n} a_{ii}$.

The trace of A is a single real or complex number.

1.1.13 **Proposition**

The trace has the following properties.

(i) trace $(A + B)$ = trace A + trace B for all n×n matrices A and B.

(ii) trace $(\lambda A) = \lambda$ trace A for all n×n matrices A and all scalars λ.

(iii) trace A^t = trace A for all nxn matrices A.

(iv) trace AB = trace BA for all nxn matrices A and B.

Proof

Easy exercise to prove (i),(ii), and (iii). To prove (iv) note that the (i,i)-entry of AB is $\sum_{j=1}^{n} a_{ij}b_{ji}$ which yields that trace AB $= \sum_{i=1}^{n}\sum_{j=1}^{n} a_{ij}b_{ji}$. Since both i and j are being summed from 1 to n this last double sum is symmetric in A and B and thus it must also give the value of trace BA.

Problems 1A

1. Let $A = \begin{pmatrix} 2 & 3 \\ 1 & 2 \end{pmatrix}$, $B = \begin{pmatrix} 4 & -1 \\ 4 & 0 \end{pmatrix}$, $C = \begin{pmatrix} -1 & 2 \\ 2 & -1 \\ 1 & 3 \end{pmatrix}$, $D = \begin{pmatrix} 3 & 2 & 1 \\ 4 & -6 & 0 \\ 1 & -2 & -2 \end{pmatrix}$.

Calculate each of the following matrix products;

$$AB,\ CA,\ DC,\ DCAB,\ A^2,\ D^2,\ A^3B^2$$

2. Let $A = \begin{pmatrix} 1 & 1 & 0 \\ 0 & 1 & 2 \\ 0 & 0 & 1 \end{pmatrix}$. Prove by induction that $A^n = \begin{pmatrix} 1 & n & n(n-1) \\ 0 & 1 & 2n \\ 0 & 0 & 1 \end{pmatrix}$.

3. Let A be an mxn matrix and B an nxp matrix. Let $B_1, B_2, \ldots, B_p$ denote the columns of B. Show that $AB_1, AB_2, \ldots, AB_p$ are the columns of AB. If $A_1, A_2, \ldots, A_m$ denote the rows of A show that $A_1B, A_2B, \ldots, A_mB$ are the rows of AB.

4. Let A be an nxn matrix with entries in F. If AB = BA for all nxn matrices B with entries in F show that $A = \alpha I_n$ for some $\alpha \in F$, i.e. A is a scalar multiple of the identity matrix.

5. Let A be an nxn matrix with complex entries. If trace $\bar{A}^tA = 0$ show that A is the zero matrix.

(*Hint* - show that trace $\bar{A}^tA = \sum_{i=1}^{n}\sum_{j=1}^{n} |a_{ij}|^2$ where $|z|$ denotes the modulus of the complex number z.)

6. Let E_{ij} denote the nxn matrix with entry 1 in the (i,j)-place and zero elsewhere. Show that any nxn matrix $A = (a_{ij})$ is expressible in the form

$$A = \sum_{i=1}^{n} \sum_{j=1}^{n} a_{ij} E_{ij}.$$

Show also that $E_{ij} E_{kl} = 0$ if $j \neq k$, and $E_{ij} E_{jl} = E_{il}$.

7. Let the n×n matrix X be partitioned as follows ;

$X = \begin{pmatrix} A & B \\ C & D \end{pmatrix}$ where A is a p×p matrix, B is a p×q matrix, C is a q×p matrix, and D is a q×q matrix where $p + q = n$.

Let $Y = \begin{pmatrix} E & F \\ G & H \end{pmatrix}$ be an n×n matrix partitioned in a similar way. (i.e. E is a p×p matrix etc.)

Show that the product XY is partitioned as follows.

$$XY = \begin{pmatrix} AE + BG & AF + BH \\ CE + DG & CF + DH \end{pmatrix}$$

1.2 Systems of linear equations

A system of simultaneous linear equations

$$\begin{aligned} a_{11}x_1 + a_{12}x_2 + \ldots\ldots + a_{1n}x_n &= b_1 \\ a_{21}x_1 + a_{22}x_2 + \ldots\ldots + a_{2n}x_n &= b_2 \\ &\vdots \\ a_{m1}x_1 + a_{m2}x_2 + \ldots\ldots + a_{mn}x_n &= b_m \end{aligned}$$

in n unknowns $x_1, x_2, \ldots, x_n$ can be rewritten as a single matrix equation $Ax = b$ where $A = (a_{ij})$ is an m×n matrix, $b = (b_i)$ is a column vector of length m, and $x = (x_i)$ is a column vector of length n.

We assume that the entries of A and b are real.

A *solution* of the system is an n-tuple of real numbers $(\alpha_1, \alpha_2, \ldots \alpha_n)$ such that $x_i = \alpha_i$ for each $i = 1,2,\ldots,n$ satisfies each of the m equations.

The *solution set* of the system is the set of all solutions of the system. It is a subset of $\mathbb{R}^n$. There are three possibilities for the solution set of the system;

(i) there is a unique solution, i.e. the solution set consists of a single point,

(ii) there are infinitely many solutions,

(iii) there are no solutions at all, i.e. the solution set is empty.

(In this case we say that the equations are *inconsistent*.)

For $m < n$ only possibilities (ii) and (iii) can occur whereas for $m \geq n$ all three possibilities can occur.

We illustrate this with a few simple examples;

1.2.1 **Example**

$$2x_1 + 3x_2 = 8$$
$$3x_1 - 3x_2 = 2$$

This system of two equations in two unknowns has the unique solution $x_1 = 2$, $x_2 = 4/3$.

Geometrically the two equations each represent a line in the plane and the solution set of the system is the point of intersection of the two lines.

1.2.2 **Example**

$$2x_1 + 3x_2 = 8$$
$$4x_1 + 6x_2 = 16$$

This system of two equations in two unknowns has infinitely many solutions. Specifically $x_1 = \alpha$, $x_2 = (8 - 2\alpha)/3$ for any $\alpha \in \mathbb{R}$ will be a solution.

Geometrically the two equations each represent the same line in the plane and the solution set of the system is the infinite set of all points on this line.

1.2.3 **Example**

$$2x_1 + 3x_2 = 8$$
$$4x_1 + 6x_2 = 3$$

This system of two equations in two unknowns has no solutions, the two equations being inconsistent.

Geometrically the two equations represent two parallel lines and so there are no points common to the two lines.

1.2.4 **Example**

$$x_1 + 2x_2 + x_3 = 3$$
$$x_1 - x_2 - x_3 = 2$$

Adding these two equations yields $2x_1 + x_2 = 5$. This gives $x_2 = 5 - 2x_1$. Substituting into the first equation of the system then gives

$x_3 = 3 - x_1 - 2x_2 = 3 - x_1 - 2(5 - 2x_1) = 3x_1 - 7$.

Thus x_1 is free to take any real number value and x_2 and x_3 are then given in terms of x_1.

The solution set is $\{ (\alpha, 5 - 2\alpha, 3\alpha - 7) ; \alpha \in \mathbb{R} \}$.

Geometrically the two equations of the system each represent a plane in $\mathbb{R}^3$ and the solution set is the line of intersection of the two planes.

1.2.5 **Remark**

In this last example $m = 2$, $n = 3$, i.e. there are more unknowns than equations. In that situation a unique solution to the system cannot be expected. There is insufficient information to be able to obtain a unique value for the unknowns so that possibility (i) cannot occur.

Geometrically two equations in three unknowns represent two planes in $\mathbb{R}^3$. These two planes can either intersect in a line as in example (1.2.4) or else be parallel and so have no points of intersection, i.e. the solution set of the corresponding system is the empty set. Similar geometric considerations preclude the possibibility of there being a unique solution in the case of a system of m equations in n unknowns with $m < n$ for larger values of m and n.

1.2.6 **Some special kinds of system**

An mxn matrix is called an *echelon matrix* if it satisfies the following two conditions;

(i) the first non-zero entry in each row is one,

(ii) for any two consecutive rows either the lower of the two rows contains only zeros or else the first non-zero entry of the lower row is to the right of the first non-zero entry of the upper row.

Below are some examples of echelon matrices;

$$\begin{pmatrix} 1 & 3 & 2 \\ 0 & 1 & 1 \\ 0 & 0 & 1 \end{pmatrix}, \quad \begin{pmatrix} 1 & 1 & 3 & -1 \\ 0 & 1 & 2 & 1 \\ 0 & 0 & 1 & 3 \end{pmatrix}, \quad \begin{pmatrix} 1 & 3 & 4 & 1 \\ 0 & 0 & 1 & 2 \\ 0 & 0 & 0 & 0 \\ 0 & 0 & 0 & 0 \end{pmatrix}, \quad \begin{pmatrix} 1 & 2 & 1 & 2 \\ 0 & 0 & 0 & 1 \\ 0 & 0 & 0 & 0 \\ 0 & 0 & 0 & 0 \end{pmatrix}$$

When A is an echelon matrix the system of equations $Ax = b$ is very easily solved by a method called *back substitution.* We illustrate this method by a few examples;

1.2.7 **Example**

$$\begin{pmatrix} 1 & 3 & 2 \\ 0 & 1 & 1 \\ 0 & 0 & 1 \end{pmatrix} \begin{pmatrix} x_1 \\ x_2 \\ x_3 \end{pmatrix} = \begin{pmatrix} 3 \\ 4 \\ 5 \end{pmatrix}$$

Our equations are $x_1 + 3x_2 + 2x_3 = 3$

$$x_2 + x_3 = 4$$

$$x_3 = 5$$

Starting with the last equation $x_3 = 5$, we substitute into the equation above to obtain $x_2 = -1$, and then substitute into the equation above that to obtain $x_1 = -4$. This system thus has the unique solution $x_1 = -4$, $x_2 = -1$, $x_3 = 5$.

1.2.8 **Example**

$$\begin{pmatrix}1&2&1&2\\0&0&0&1\\0&0&0&0\\0&0&0&0\end{pmatrix}\begin{pmatrix}x_1\\x_2\\x_3\\x_4\end{pmatrix}=\begin{pmatrix}4\\2\\0\\0\end{pmatrix}$$

Our equations are $x_1 + 2x_2 + x_3 + 2x_4 = 4$,

$$x_4 = 2.$$

Substituting $x_4 = 2$ into the equation above yields that $x_1 + 2x_2 + x_3 = 0$. Thus two of the variables x_1,x_2,x_3 can take any real number value and the third one is then determined in terms of these. If we take $x_1 = \alpha$, $x_2 = \beta$ where $\alpha \in \mathbb{R}$, $\beta \in \mathbb{R}$, then we can write $x_3 = -\alpha - 2\beta$. Our solution set may thus be written in the form $\{(\alpha,\beta,-\alpha - 2\beta,2);\ \alpha \in \mathbb{R},\beta \in \mathbb{R}\}$.

1.2.9 **Example**

$$\begin{pmatrix}1&3&4&1\\0&0&1&2\\0&0&0&0\\0&0&0&0\end{pmatrix}\begin{pmatrix}x_1\\x_2\\x_3\\x_4\end{pmatrix}=\begin{pmatrix}5\\8\\3\\0\end{pmatrix}$$

Our equations are $x_1 + 3x_2 + 4x_3 + x_4 = 5$,

$$x_3 + 2x_4 = 8,$$

$$0 = 3.$$

Clearly the last equation is impossible and so the system has no solution, i.e. the solution set is the empty set.

1.2.10 **Remark**

The solution sets in some of the above examples contain *free variables*, i.e. variables which may take any real number value, the other variables being given in terms of these free variables. In example (1.2.8) there are two free variables , in example (1.2.4) there is one free variable.

The number of free variables in the solution set is called the *dimension* of the solution set. (It is also sometimes called the *number of degrees of freedom.*) If there is a unique solution, i.e. the solution set is a single point of $\mathbb{R}^n$, then the dimension of the solution set is said to be zero.

1.2.11 **Elementary operations**

The following three kinds of operation on a set of simultaneous linear equations are called *elementary operations*;

(1) Interchange of a pair of equations.

(2) Multiplication of one equation by a non-zero scalar λ.

(3) Addition of λ times one equation to another equation, λ being a scalar.

1.2.12 **Definition**

Two systems of m simultaneous linear equations in n unknowns are *equivalent* if one system is obtainable from the other by a finite sequence of elementary operations.

1.2.13 **Proposition**

Two equivalent systems of simultaneous linear equations will have exactly the same solution set.

Proof

If the system $Cx = d$ is obtained from the system $Ax = b$ by a single elementary operation then any solution of $Ax = b$ will also be a solution of $Cx = d$. Since each of the three kinds of elementary operation is reversible it follows that $Ax = b$ and $Cx = d$ will have exactly the same solution set. The result follows immediately.

1.2.14 **Remark**

Our approach for solving a system of linear equations will be to transform the system into an equivalent one of the form $Cx = d$ where C is an echelon matrix. This system is then solved by back substitution.

1.2.15 The augmented matrix

Let $Ax = b$ be a system of m linear equations in n unknowns. The $(m + 1)\times n$ matrix $[A,b]$ obtained by adjoining the column b to the $m\times n$ matrix A is called the *augmented matrix* of the system.

1.2.16 Elementary row operations

The following three kinds of operation on a matrix are called *elementary row operations*;

(1) Interchange of two rows of a matrix.

(2) Multiplication of one row of the matrix by a non-zero scalar λ

(3) Addition of λ times one row of the matrix to another row, λ being a scalar.

Performing the elementary row operations of kind (1),(2), or (3) on the augmented matrix $[A,b]$ of a system of equations $Ax = b$ corresponds exactly to performing the elementary operations of (1.2.11) on the set of equations.

1.2.17 Comment

Each operation in (1.2.16) amounts to left multiplication by a certain matrix.

An operation of kind (1) where row r and row s are interchanged amounts to left multiplication by the $m\times m$ matrix E which has entries $e_{ii} = 1$ for all $i \neq r$, $i \neq s$, $e_{rs} = e_{sr} = 1$ and $e_{ij} = 0$ otherwise.

For $m = 3$, $r = 1$, $s = 3$, $E = \begin{pmatrix} 0 & 0 & 1 \\ 0 & 1 & 0 \\ 1 & 0 & 0 \end{pmatrix}$.

An operation of kind (2) where row r is multiplied by λ amounts to left multiplication by the $m\times m$ matrix E which has entries $e_{ii} = \lambda$ for $i = r$, $e_{ii} = 1$ for all $i \neq r$, and $e_{ij} = 0$ otherwise.

For $m = 3$, $r = 2$, $E = \begin{pmatrix} 1 & 0 & 0 \\ 0 & \lambda & 0 \\ 0 & 0 & 1 \end{pmatrix}$.

An operation of kind (3) where λ times row s is added to row r amounts to left multiplication by the matrix E with entries $e_{ii} = 1$ for all i, $e_{rs} = \lambda$, and $e_{ij} = 0$ otherwise.

For m = 3, r = 1, s = 2, $E = \begin{pmatrix} 1 & \lambda & 0 \\ 0 & 1 & 0 \\ 0 & 0 & 1 \end{pmatrix}$.

The above three kinds of matrices E are known as *elementary matrices.*

1.2.18 **Gaussian elimination**

To solve a system of simultaneous linear equations Ax = b we perform a sequence of elementary row operations on the augmented matrix [A,b] to obtain [C,d] where C is an echelon matrix. It is not difficult to see that any m×n matrix A can be reduced to an echelon matrix by a sequence of elementary row operations. The argument goes as follows.

We may assume that the first column of A does not consist entirely of zeros since if it did the first variable would be redundant. By interchanging rows if necessary and multiplying by a suitable scalar we can ensure that 1 appears in the (1,1)-place. Then subtracting suitable multiples of row 1 from the other rows makes all other entries in column 1 equal to zero. Now look at the (n-1)×(n-1) matrix obtained by omitting row 1 and column 1. If the first column of this (n-1)×(n-1) matrix does not consist entirely of zeros we can, in the same manner as above, make the (1,1)-entry of this matrix equal to one and all the entries below it equal to zero. If the first column of the (n-1)×(n-1) matrix does consist entirely of zeros move to the second column. If this column does not consist entirely of zeros we can, as above, make it into a column with 1 on top and all zeros below. If the column does consist entirely of zeros move to the third column. Proceeding in this fashion we eventually must obtain an echelon form.

By (1.2.13) the solution set of the system Ax = b will be identical to the solution set of Cx = d and since C is an echelon matrix we can easily find its solution set by back substitution.

The method we have described is called *the method of Gaussian elimination.*

We illustrate the method by a couple of examples;

1.2.19 **Example**

$$x_1 + 2x_2 + x_3 - x_4 = 0$$
$$2x_2 + 3x_3 + 3x_4 = 8$$
$$x_1 - x_2 - 3x_3 - 4x_4 = -8$$
$$x_1 + x_2 + 5x_3 - 2x_4 = -8$$

We write the augmented matrix as follows;

$$\left(\begin{array}{cccc|c} 1 & 2 & 1 & -1 & 0 \\ 0 & 2 & 3 & 3 & 8 \\ 1 & -1 & -3 & -4 & -8 \\ 1 & 1 & 5 & -2 & -8 \end{array}\right)$$

Subtracting row 1 from rows 3 and 4 yields

$$\left(\begin{array}{cccc|c} 1 & 2 & 1 & -1 & 0 \\ 0 & 2 & 3 & 3 & 8 \\ 0 & -3 & -4 & -3 & -8 \\ 0 & -1 & 4 & -1 & -8 \end{array}\right)$$

Adding twice row 4 to row 2 and subtracting three times row 4 from row 3 yields

$$\left(\begin{array}{cccc|c} 1 & 2 & 1 & -1 & 0 \\ 0 & 0 & 11 & 1 & -8 \\ 0 & 0 & -16 & 0 & 16 \\ 0 & -1 & 4 & -1 & -8 \end{array}\right)$$

Dividing row 3 by -16, subtracting 11 times the new row 3 from row 2, multiplying row 4 by -1 ,and interchanging rows yields the required echelon form.

$$\left(\begin{array}{cccc|c} 1 & 2 & 1 & -1 & 0 \\ 0 & 1 & -4 & 1 & 8 \\ 0 & 0 & 1 & 0 & -1 \\ 0 & 0 & 0 & 1 & 3 \end{array}\right)$$

Back substitution quickly gives the unique solution $x_4 = 3$, $x_3 = -1$, $x_2 = 1$, $x_1 = 2$.

1.2.20 **Example**

$$2x_1 - x_2 - x_3 = 2$$

$$x_1 + x_3 - 4x_4 = 1$$

$$x_2 - x_3 - 4x_4 = 4$$

The augmented matrix for this system is as follows;

$$\left(\begin{array}{cccc|c} 2 & -1 & -1 & 0 & 2 \\ 1 & 0 & 1 & -4 & 1 \\ 0 & 1 & -1 & -4 & 4 \end{array}\right)$$

Subtracting twice row 2 from row 1 yields

$$\left(\begin{array}{cccc|c} 0 & -1 & -3 & 8 & 0 \\ 1 & 0 & 1 & -4 & 1 \\ 0 & 1 & -1 & -4 & 4 \end{array}\right)$$

Adding row 3 to row 1 yields

$$\left(\begin{array}{cccc|c} 0 & 0 & -4 & 4 & 4 \\ 1 & 0 & -1 & -4 & 1 \\ 0 & 1 & -1 & -4 & 4 \end{array}\right)$$

Dividing row 1 by -4 and interchanging rows yields the desired echelon form

$$\left(\begin{array}{cccc|c} 1 & 0 & 1 & -4 & 1 \\ 0 & 1 & -1 & -4 & 4 \\ 0 & 0 & 1 & -1 & -1 \end{array}\right)$$

Back substitution quickly shows that there is one free variable and the solution set is $x_4 = \alpha$, $x_3 = \alpha - 1$, $x_2 = 5\alpha + 3$, $x_1 = 3\alpha + 2$, where $\alpha \in \mathbb{R}$.

1.2.21 **Comment**

Let C be an echelon matrix. If in the system $Cx = d$ we find that one row, say row r, of C consists entirely of zeros while the corresponding entry d_r of d is non-zero then the system will have no solution, i.e. the equations are inconsistent.

The method of Gaussian elimination we have described is equivalent to operating directly on the set of equations and successively eliminating variables. For computational purposes it is better to operate on the augmented matrix rather than on the equations themselves. When the number of equations and unknowns is large the procedure would be implemented on a computer.

1.2.22 Comment

We stated earlier that the coefficients in our systems of equations belonged to the field $\mathbb{R}$ of real numbers. In fact the above method for solving a system of linear equations works equally well if the coefficients belong to the field $\mathbb{C}$ of complex numbers or to the field $\mathbb{Q}$ of rational numbers. The solutions will of course then have values in $\mathbb{C}$ or $\mathbb{Q}$ respectively.

It should be noted that if all of the coefficients of a system are integers then the solutions will be rational numbers but not necessarily integers since fractions may occur in the reduction to echelon form.

Problems 1B

1. Solve by back substitution each of the following ;

(i)

$$\begin{aligned} 3x_1 + x_2 + 2x_3 - x_4 &= 2 \\ 2x_2 + 3x_3 + 4x_4 &= -2 \\ x_3 - 6x_4 &= 9 \\ x_4 &= 4 \end{aligned}$$

(ii)

$$\begin{aligned} 3w + 2x + 2y - z &= 2 \\ 2x + 3y + 4z &= -2 \\ y - 6z &= 6 \end{aligned}$$

2. Use Gaussian elimination to solve

$$\begin{pmatrix} 1 & 4 & 1 & 3 \\ 0 & 2 & 1 & 3 \\ 1 & 3 & 1 & 2 \\ 0 & 2 & 1 & 6 \end{pmatrix} \begin{pmatrix} w \\ x \\ y \\ z \end{pmatrix} = \begin{pmatrix} 1 \\ 0 \\ 1 \\ 0 \end{pmatrix}$$

3. Solve each of the following ;

(a)

$$w - x + y - z = 1$$

$$w + x - y - z = 1$$

$$w - x - y + z = 2$$

$$4w - 2x - 2y = 1$$

(b)

$$x + y + z = 1$$

$$x - y + 2z = 1$$

$$2x + 3z = 2$$

$$2x + 6y = 2$$

4. Show that the following system has a unique solution for all $k \neq -1$ but has no solution for $k = -1$.

$$x + y + kz = 1$$

$$x - y - z = 2$$

$$2x + y - 2z = 3$$

5. Solve each of the following ;

(a)

$$x - y - z = 1$$

$$2x - y + 2z = 7$$

$$x + y + z = 5$$

$$x - 2y - z = 0$$

(b)

$$x + 2y - z = 2$$

$$2x - y - 2z = 4$$

$$x + 12y - z = 2$$

6. Let A be an $m \times n$ matrix, x a column vector of length n, and let O denote the column vector of length m consisting entirely of zeros. Show that the system $Ax = O$ is always consistent, i.e. the solution set is non-empty. (A system of the form $Ax = O$ is called a *homogeneous system.*)

If the vector v satisfies $Av = b$ show that the vector $v + w$ will be a solution of $Ax = b$ whenever w is a solution of the homogeneous system $Ax = 0$ Show that every solution of $Ax = b$ is expressible in this form.

1.3 The inverse of a square matrix

Let A be an nxn matrix. If there exists an nxn matrix B such that $AB = BA = I_n$, the nxn identity matrix, then B is said to be an *inverse* of A.

1.3.1 Remark

If B_1 and B_2 are each inverses of A then $B_1 = B_2$. This is because $B_2 = B_2(AB_1) = (B_2A)B_1 = B_1$. Thus if an inverse of A exists it is unique and so we may talk about *the* inverse of A.

1.3.2 Notation and terminology

The inverse of A is denoted A^{-1}. We say that A is *non-singular* (or *invertible)* if the inverse of A exists. Otherwise we say that A is *singular.*

1.3.3 Proposition

The inverse has the following properties;

(i) $(A^{-1})^{-1} = A$ for all invertible nxn matrices A.

(ii) $(AB)^{-1} = B^{-1}A^{-1}$ for all invertible nxn matrices A and B.

(iii) $(A^t)^{-1} = (A^{-1})^t$ for all invertible nxn matrices A.

Proof

Easy exercise.

1.3.4 Lemma

Each of the three kinds of elementary matrix in (1.2.17) is invertible.

Proof

The inverse of an elementary matrix will be an elementary matrix of the same kind.

For an elementary matrix E of the first kind $E^2 = I$ and so $E^{-1} = E$. For an elementary matrix E of the second kind which has the non-zero scalar λ in the (i,i)-place E^{-1} will have λ^{-1} in the (i,i)-place.

For an elementary matrix of the third kind which has the scalar λ in the (r,s)-place E^{-1} will have $-\lambda$ in the (r,s)-place.

1.3.5 **Proposition**

Let A be an $n \times n$ matrix. The following three statements about A are equivalent ;

(i) A is invertible.

(ii) The system of linear equations $Ax = b$ has a unique solution for some vector b.

(iii) A is expressible as a product of elementary matrices.

Proof

If A is invertible then the system $Ax = b$ has the unique solution $x = A^{-1}b$. Thus (i) implies (ii).

To prove that (ii) implies (iii) note first that if the system of equations $Ax = b$ has a unique solution then A must reduce to an echelon form with entries 1 at each point on the diagonal. (The *diagonal* of a square matrix is the set of all (i,i)-entries.) By a further set of elementary row operations we may reduce this echelon form to the identity matrix. Hence if the system $Ax = b$ has a unique solution then there exist elementary matrices $E_1, E_2, \ldots E_r$ such that $E_1 E_2 \cdot\,.\,.\,. E_r A = I$. Multiply this equation on the left successively by $E_1^{-1}, E_2^{-1}, \ldots, E_r^{-1}$ and we obtain A as a product of elementary matrices by (1.3.4).

Suppose $A = E_1 E_2 \cdot\,.\,. E_r$ where E_i, $i = 1,2,\ldots,r$ are elementary matrices. Then A is invertible by (1.3.3)(ii) and (1.3.4). This shows that (iii) implies (i).

1.3.6 **Remark**

If the system $Ax = b$ has a unique solution for some vector b then it has a unique solution for every vector b. (The unique solution is of course given by $x = A^{-1}b$.)

1.3.7 Calculation of the inverse

Proposition (1.3.5) yields an effective way to calculate A^{-1}. Perform a sequence of elementary operations on A to reduce it to the identity matrix I. Then performing exactly the same sequence on I will yield A^{-1}. The following example illustrates this ;

1.3.8 Example

$$A = \begin{pmatrix} 1 & 3 & 1 \\ -1 & 1 & 2 \\ 2 & 1 & -2 \end{pmatrix}$$

We write A and I alongside each other and perform the elementary operations on them together.

$$\left.\begin{pmatrix} 1 & 3 & 1 \\ -1 & 1 & 2 \\ 2 & 1 & -2 \end{pmatrix}\right| \begin{pmatrix} 1 & 0 & 0 \\ 0 & 1 & 0 \\ 0 & 0 & 1 \end{pmatrix}$$

$$\left.\begin{pmatrix} 1 & 3 & 1 \\ 0 & 4 & 3 \\ 0 & -5 & -4 \end{pmatrix}\right| \begin{pmatrix} 1 & 0 & 0 \\ 1 & 1 & 0 \\ -2 & 0 & 1 \end{pmatrix}$$

$$\left.\begin{pmatrix} 1 & 3 & 1 \\ 0 & 20 & 15 \\ 0 & -20 & -16 \end{pmatrix}\right| \begin{pmatrix} 1 & 1 & 0 \\ 5 & 5 & 0 \\ -8 & 0 & 4 \end{pmatrix}$$

$$\left.\begin{pmatrix} 1 & 3 & 1 \\ 0 & 20 & 15 \\ 0 & 0 & -1 \end{pmatrix}\right| \begin{pmatrix} 1 & 0 & 0 \\ 5 & 5 & 0 \\ -3 & 5 & 4 \end{pmatrix}$$

$$\left.\begin{pmatrix} 1 & 3 & 1 \\ 0 & 4 & 3 \\ 0 & 0 & 1 \end{pmatrix}\right| \begin{pmatrix} 1 & 0 & 0 \\ 1 & 1 & 0 \\ 3 & -5 & -4 \end{pmatrix}$$

$$\left.\begin{pmatrix} 1 & 3 & 0 \\ 0 & 4 & 0 \\ 0 & 0 & 1 \end{pmatrix}\right| \begin{pmatrix} -2 & 5 & 4 \\ -8 & 16 & 12 \\ 3 & -5 & -4 \end{pmatrix}$$

$$\left.\begin{pmatrix} 1 & 3 & 0 \\ 0 & 1 & 0 \\ 0 & 0 & 1 \end{pmatrix}\right| \begin{pmatrix} -2 & 5 & 4 \\ -2 & 4 & 3 \\ 3 & -5 & -4 \end{pmatrix}$$

$$\left(\begin{matrix}1&0&0\\0&1&0\\0&0&1\end{matrix}\;\middle|\;\begin{matrix}4&-7&-5\\-2&4&3\\3&-5&-4\end{matrix}\right)$$

We performed the following sequence of operations;

Add row 1 to row 2 and subtract twice row 1 from row 3

Multiply row 2 by 5 and multiply row 3 by 4

Add row 2 to row 3

Divide row 2 by 5

Subtract row 3 from row 1 and subtract 3 times row 3 from row 2

Divide row 2 by 4

Subtract 3 times row 2 from row 1.

We have thus found that $A^{-1} = \begin{pmatrix}4&-7&-5\\-2&4&3\\3&-5&-4\end{pmatrix}$.

As a check on the calculation of A^{-1} the reader should always verify that $AA^{-1} = I$.

1.3.9 **Comment**

If we replace the word "row" by the word "column" everywhere in (1.2.16) we can define *elementary column operations.* Performing an elementary column operation on a matrix A amounts to multiplying A on the **right** by an elementary matrix. Elementary row operations occurred naturally in (1.2) in the manipulation of systems of linear equations whereas column operations would not have been appropriate there. However all we have said in this section (1.3) on calculating inverses can be done equally well using elementary column operations instead of row operations. It is important though not to mix row and column operations, i.e. to find A^{-1} we either reduce A to the identity matrix I using only elementary row operations or else reduce A to I using only elementary column operations.

1.4 Determinants

For each square matrix A with entries in F, F = ℝ or ℂ, we can associate a single element of F called the *determinant* of A and denoted det A for short.

1.4.1 Definition

Let $A = (a_{11})$ be a 1×1 matrix. We define $\det A = a_{11}$.

Let $A = (a_{ij})$ be a 2×2 matrix. We define $\det A = a_{11}a_{22} - a_{12}a_{21}$.

Now assume that the determinant has been defined for all $(n - 1) \times (n - 1)$ matrices and let $A = (a_{ij})$ be an $n \times n$ matrix.

We denote by M_{ij} the determinant of the $(n - 1) \times (n - 1)$ matrix obtained from A by deleting row i and column j of A. We call M_{ij} the *minor* of A corresponding to the entry a_{ij} of A.

We now define the determinant of the $n \times n$ matrix A by

$$\det A = \sum_{j=1}^{n} (-1)^{j+1} a_{1j} M_{1j}.$$

Note that for n = 2 this reduces to the definition already given and for n = 3 we have the formula

$$\det A = a_{11} \det \begin{pmatrix} a_{22} & a_{23} \\ a_{32} & a_{33} \end{pmatrix} - a_{12} \det \begin{pmatrix} a_{21} & a_{23} \\ a_{31} & a_{33} \end{pmatrix} + a_{13} \det \begin{pmatrix} a_{21} & a_{22} \\ a_{31} & a_{32} \end{pmatrix}.$$

The main properties of determinants are contained in the following proposition;

1.4.2 Proposition

(1) $\det A = \sum_{j=1}^{n} (-1)^{j+1} a_{j1} M_{j1}$.

(2) $\det A = \det A^t$ for all square matrices A.

(3) If A is upper triangular, i.e. $a_{ij} = 0$ for all $i > j$,

then $\det A = a_{11}a_{22}a_{33} \ldots\ldots a_{nn}$.

If A is lower triangular, i.e. $a_{ij} = 0$ for all $i < j$, then det A = $a_{11}a_{22}a_{33}......a_{nn}$.

(4) If one row of A consists entirely of zeros then det A = 0.

(5) If B is the matrix obtained from A by multiplying one row of A by the scalar λ then det B = λ det A.

(6) If B is the matrix obtained from A by interchanging two rows of A then det B = -det A.

(7) If two rows of A are identical then det A = 0.

(8) If B is the matrix obtained from A by adding λ times one row of A to another row of A then det B = det A.

(9) Properties (4), (5), (6), (7),and (8) remain valid if the word "row" is replaced everywhere by the word "column".

(10) det AB = (det A)(det B) for any pair of n$\times$n matrices A and B.

Proof

The proof of all these properties is rather long and is left until the appendix to this chapter.

1.4.3 **Remark**

The formula in our definition of det A is often called *the expansion along the first row* because the entries a_{1j}, j = 1,2,...n of row 1 appear in the formula.

Property (1) shows that det A can equivalently be obtained via an *expansion down the first column.*

Evaluation of det A from the basic definition involves a lot of calculation even if n is as small as 4. It is easier to reduce A to upper or lower triangular form by row operations and utilize the properties given in (1.4.2). We illustrate this by the following example;

1.4.4 **Example**

Find det A where $A = \begin{pmatrix} 3 & 1 & 1 & 2 \\ 1 & 2 & 0 & 1 \\ 1 & 1 & 2 & -1 \\ -2 & 1 & -1 & 3 \end{pmatrix}$.

$$\det A = \det \begin{pmatrix} 3 & 1 & 1 & 2 \\ 1 & 2 & 0 & 1 \\ 1 & 1 & 2 & -1 \\ -2 & 1 & -1 & 3 \end{pmatrix} = \det \begin{pmatrix} 0 & -5 & 1 & -1 \\ 1 & 2 & 0 & 1 \\ 0 & -1 & 2 & -2 \\ 0 & 5 & -1 & 5 \end{pmatrix}$$

after subtracting 3 times row 2 from row 1, subtracting row 2 from row 3, and adding twice row 2 to row 4. This uses property (8).

Adding row 1 to row 4 and interchanging row 1 and row 2 then yields , using properties (6) and (8), that

$$\det A = -\begin{pmatrix} 1 & 2 & 0 & 1 \\ 0 & -5 & 1 & -1 \\ 0 & -1 & 2 & -2 \\ 0 & 0 & 0 & 4 \end{pmatrix}$$

Subtracting 5 times row 3 from row 2 and then interchanging row 2 with row 3 yields, using (8) and (6) again, that

$$\det A = \begin{pmatrix} 1 & 2 & 0 & 1 \\ 0 & -1 & 2 & -2 \\ 0 & 0 & -9 & 9 \\ 0 & 0 & 0 & 4 \end{pmatrix}$$

Property (3) now gives det A = 36.

1.4.5 **Remark**

It is possible to give a more sophisticated definition of the determinant and some textbooks do this. Their definition is equivalent to ours. However the definition of (1.4.1) and knowledge of the properties of (1.4.2) are sufficient for our purposes.

1.4.6 **The adjugate of a matrix**

Let $A = (a_{ij})$ be an nxn matrix. The *cofactor* C_{ij} of A corresponding to the entry a_{ij} is defined by $C_{ij} = (-1)^{i+j}M_{ij}$ where M_{ij} is the minor as defined in (1.4.1).

Note that $\det A = \sum_{j=1}^{n} a_{1j}C_{1j}$.

The *adjugate of A*, denoted adj A, is the n×n matrix which has entry C_{ji} in the (i,j)-place. Thus adj A is the transpose of the matrix of cofactors of A.

To obtain adj A we first find the matrix of minors M_{ij}, then multiply alternate entries of this matrix by -1, and finally transpose the matrix. The following diagram may be helpful in remembering where to multiply by -1.

$$\begin{pmatrix} + & - & + & - & . & . \\ - & + & - & + & . & \\ + & - & + & - & . & \\ - & + & - & + & . & \\ . & . & . & . & . & \\ . & & & & & \end{pmatrix}$$

For example if $A = \begin{pmatrix} a & b \\ c & d \end{pmatrix}$ then adj $A = \begin{pmatrix} d & -b \\ -c & a \end{pmatrix}$.

1.4.7 **Proposition**

Let A be an n×n matrix and I the identity n ×n matrix. Then

A(adj A) = (adj A)A = (det A)I

Proof

See the appendix to this chapter.

1.4.8 **Proposition**

The n×n matrix A is invertible if and only if det A $\neq$ 0.

Proof

If det A $\neq$ 0 then A^{-1} exists and equals $(\det A)^{-1}$(adj A) by (1.4.7).

Conversely if A^{-1} exists then property (10) of (1.4.2) applied with $B = A^{-1}$ implies that (det A)(det A^{-1}) = det I = 1 and in particular det A cannot be zero.

1.4.9 **Calculation of the inverse via the adjugate**

We have seen in (1.3.7) and (1.3.8) one method of calculating A^{-1}. Proposition (1.4.7) yields a different method which is very quick for 2×2 and 3x3 matrices. For larger matrices the calculations become very cumbersome and the method of (1.3.7) for finding inverses is more efficient from a computational viewpoint.

Let $A = \begin{pmatrix} a & b \\ c & d \end{pmatrix}$. Then A^{-1} exists provided $ad - bc \neq 0$ and

$$A^{-1} = (ad - bc)^{-1}\begin{pmatrix} d & -b \\ -c & a \end{pmatrix}.$$

1.4.10 **Example**

Show that $A = \begin{pmatrix} 1 & 0 & 2 \\ 2 & 3 & 1 \\ 1 & 2 & 1 \end{pmatrix}$ is invertible and find A^{-1}.

The matrix of minors of A is $\begin{pmatrix} 1 & 1 & 1 \\ -4 & 1 & 2 \\ -6 & 3 & 3 \end{pmatrix}$ and det A = 3.

Hence $A^{-1} = (1/3)\text{adj } A = (1/3)\begin{pmatrix} 1 & 4 & -6 \\ -1 & -1 & 3 \\ 1 & -2 & 3 \end{pmatrix}$.

1.4.11 **Exercise**

For the matrix A of Example (1.3.8) verify that the above method yields the same value for A^{-1} as was obtained in (1.3.8).

Problems 1C

1. Use the method of 1.3 to find the inverse of each of the following;

$$\begin{pmatrix} -1 & 2 & 1 \\ 2 & 1 & -1 \\ 1 & 9 & 1 \end{pmatrix}, \quad \begin{pmatrix} 1 & 2 & 3 & 4 \\ 0 & 1 & 2 & 3 \\ 0 & 0 & 1 & 2 \\ 0 & 0 & 0 & 1 \end{pmatrix}, \quad \begin{pmatrix} 2 & 2 & 2 \\ 0 & 2 & 2 \\ 0 & 0 & 2 \end{pmatrix}.$$

2. Let $X = \begin{pmatrix} 0 & 1 & 0 \\ 0 & 0 & 1 \\ -a & -b & -c \end{pmatrix}$. Show that $aI + bX + cX^2 = O$. Deduce that X is invertible if $a \neq 0$ and that $X^{-1} = -(1/a)(bI + cX)$.

3. Let $X = \begin{pmatrix} A & B \\ C & D \end{pmatrix}$ be partitioned as in problem 7 of problems 1A. If X is invertible and if all the relevant inverses exist show that

$$X^{-1} = \begin{pmatrix} (A - BD^{-1}C)^{-1} & A^{-1}B(CA^{-1}B - D)^{-1} \\ (CA^{-1}B - D)^{-1}CA^{-1} & (D - CA^{-1}B)^{-1} \end{pmatrix}.$$

4. Calculate the determinant of each of the following matrices;

$$\begin{pmatrix} 2 & 3 & 4 \\ 0 & 1 & 2 \\ 2 & 0 & 1 \end{pmatrix}, \quad \begin{pmatrix} 1 & -1 & 2 & -2 \\ 2 & -2 & 1 & -1 \\ -2 & 1 & -1 & 2 \\ -1 & 2 & -2 & 1 \end{pmatrix}, \quad \begin{pmatrix} 2 & 3 & 1 & 2 \\ 0 & 2 & 1 & 3 \\ 0 & 2 & 1 & -2 \\ -2 & 1 & -1 & 1 \end{pmatrix}.$$

5. Show that $\det \begin{pmatrix} 1 & 1 & 1 \\ a & b & c \\ a^2 & b^2 & c^2 \end{pmatrix} = (a - b)(b - c)(c - a)$.

Show that $\det \begin{pmatrix} 1 & 1 & 1 & & 1 \\ a_1 & a_2 & a_3 & & a_{n+1} \\ a_1^2 & a_2^2 & a_3^2 & & a_{n+1}^2 \\ & & & & \\ a_1^n & a_2^2 & a_3^2 & & a_{n+1}^2 \end{pmatrix} = \prod_{i<j} (a_i - a_j)$,

the product over all a_i, a_j with $i < j$.

(Matrices of the above type are called *Vandermonde matrices.*)

6. Let $A = \begin{pmatrix} a & b & c \\ c & a & b \\ b & c & a \end{pmatrix}$. Show that det A can be factorized in the form

$(a + b + c)(a + b\omega + c\omega^2)(a + b\omega^2 + c\omega)$ where $\omega = \exp(2\pi i/3)$, a primitive

cube root of unity.

Obtain a similar factorization for $\det \begin{pmatrix} a & b & c & d \\ d & a & b & c \\ c & d & a & b \\ b & c & d & a \end{pmatrix}$ which involves a

primitive fourth root of unity.

(Matrices of the above type are called *circulant matrices.*)

7. Let A,B, and C be $n \times n$ matrices and suppose that, for each $i = 2, 3, . ,n$,

we have row i of A = row i of B = row i of C.

Suppose also that row 1 of C = (row 1 of A) + (row 1 of B).

Show that det C = det A + det B.

8. Determine whether any of the following matrices are invertible. When the inverse exists use (1.4.9) to calculate it.

$$\begin{pmatrix} 1 & -1 & -2 \\ 0 & 3 & 4 \\ 0 & 0 & 2 \end{pmatrix}, \quad \begin{pmatrix} 5 & 3 & 1 \\ 8 & 1 & 2 \\ 3 & -2 & 1 \end{pmatrix}, \quad \begin{pmatrix} 1 & 0 & 2 & -2 \\ 2 & 1 & 1 & -1 \\ 1 & 0 & 0 & 2 \\ -1 & 1 & 1 & -1 \end{pmatrix}.$$

9. Let A be an $n \times n$ matrix whose entries are integers and suppose $\det A = \pm 1$. Show that all the entries of A^{-1} are integers.

10.. Let $X = \begin{pmatrix} A & B \\ C & D \end{pmatrix}$ be partitioned as in problem 7 of problems 1A.

(i) If either $B = 0$ or $C = 0$ show that $\det X = (\det A)\ (\det D)$.

(ii) Instead suppose that A is invertible. By multiplying X on the right by a suitable matrix of the form $\begin{pmatrix} I & Y \\ 0 & I \end{pmatrix}$, I being the identity matrix of the appropriate size, deduce that $\det X = (\det A)\ (\det (D - CA^{-1}B))$.

1.5 Some special types of matrices

Let $A = (a_{ij})$ be an $n \times n$ matrix with real or complex entries.

A is said to be *diagonal* if $a_{ij} = 0$ for all $i \neq j$.

A is said to be *tridiagonal* if $a_{ij} = 0$ for all i,j with $|i - j| > 1$. i.e. all the entries of A are zero except possibly for those on the main diagonal $(a_{11}, a_{22}, \ldots, a_{nn})$, the superdiagonal $(a_{12}, a_{23}, \ldots, a_{(n-1)n})$, and the subdiagonal $(a_{21}, a_{32}, \ldots, a_{n(n-1)})$.

A is said to be *upper triangular* if $a_{ij} = 0$ for all $i > j$, i.e. all entries below the main diagonal are zero.

A is said to be *lower triangular* if $a_{ij} = 0$ for all $i < j$, i.e. all entries above the main diagonal are zero.

A is said to be *upper Hessenberg* if $a_{ij} = 0$ for all $i > j + 1$, i.e. all entries above the superdiagonal are zero.

A is said to be *lower Hessenberg* if $a_{ij} = 0$ for all $j > i + 1$, i.e. all entries below the subdiagonal are zero.

A is said to be *symmetric* if $a_{ij} = a_{ji}$ for all i,j, i.e. $A^t = A$.

A is said to be *hermitian* if the entries of A are complex and $a_{ij} = \bar{a}_{ji}$ for all i,j, i.e. $\bar{A}^t = A$.

(The name hermitian comes from the French mathematician Hermite.)

A is said to be *orthogonal* if all the entries of A are real and $A^t = A^{-1}$

A is said to be *unitary* if the entries of A are complex and $\bar{A}^t = A^{-1}$.

A is said to be *block diagonal* if A is of the form

$$\begin{pmatrix} A_{11} & O & O & O \\ O & A_{22} & O & O \\ & & & \\ & & & \\ O & & O & A_{kk} \end{pmatrix}$$

where A_{ii} is an $n_i x n_i$ matrix for each i = 1,2, . . . ,k.

A is said to be a *permutation matrix* if A is obtained from the identity matrix I by a permutation of the rows of I.

Problems 1D

1. If the matrix A is symmetric and invertible show that A^{-1} is symmetric. If A is hermitian and invertible show that A^{-1} is hermitian.

2. If the matrix A is upper triangular and invertible show that A^{-1} is upper triangular. If A is lower triangular and invertible show that A^{-1} is lower triangular.

1.6 More on systems of linear equations

1.6.1 Cramer's rule

Let A be an invertible nxn matrix and Ax = b be a system of linear equations. Cramer's rule is an attractive way of describing the solution set of the system. It says that the system has a unique solution given by

$x_i = (\det A_i)/\det A$ for each $i = 1, 2, \ldots, n$ where A_i is the $n \times n$ matrix obtained from A by replacing the i-th column of A by the column vector b.

For a proof of Cramer's rule see problem 1 at the end of this section.

Cramer's rule is mainly of theoretical importance. It is not used much in practice for determining the solution of systems of linear equations because the numerical computations involved are much greater than those required for Gaussian elimination.

1.6.2 **The LU-decomposition of a square matrix**

Let A be an $n \times n$ matrix. An expression of A in the form $A = LU$ where L and U are respectively lower and upper triangular $n \times n$ matrices is called *an LU-decomposition of A*.

An LU-decomposition of A need not necessarily exist and even if it does exist it is not unique. However for any $n \times n$ matrix A there do exist permutation matrices P and Q, a lower triangular matrix L and an upper triangular matrix U such that $A = PLUQ$. Further it can be shown that if A is invertible then $A = PLU$, i.e. Q can be chosen equal to I.

Observe that if we have $A = LU$ then the system $Ax = b$ can be very easily solved in the following manner;

First solve $Ly = b$ for unknowns $y = (y_i)$ by forward elimination, (i.e. by solving equation 1 first, then equation 2 and so on). Then solve $Ux = y$ for unknowns $x = (x_i)$ by back substitution as in (1.2).

More generally if $A = PLUQ$ then we solve $LUz = P^{-1}b$ in the above manner, $z = Qx$ being a re-arrangement of the unknowns (x_i) and $P^{-1}b$ being a re-arrangement of the co-ordinates of b.

The method of Gaussian elimination as in (1.2.18) can be interpreted as yielding decompositions of the form A = LU or more generally A = PLUQ. We will explain this in more detail. Suppose first that we are able to reduce A to an echelon form U by a finite sequence of elementary row operations of kind (2) and (3) only and that the only ones of kind (3) involve adding a scalar multiple of row i to row j where i < j. Then the corresponding elementary matrices, see (1.2.17), will all be lower triangular and we have $E_rE_{r-1}........E_1A = U$ where U is an echelon matrix which is upper triangular and each E_i is an lower triangular elementary matrix.

Thus A = LU where $L = E_1^{-1}E_2^{-1}....E_r^{-1}$ is lower triangular.

L is lower triangular because the inverse of an lower triangular matrix is lower triangular, (see problem 2 of problems 1D), and the product of lower triangular matrices is lower triangular.

For a general nxn matrix A the reduction to echelon form by elementary row operations cannot be accomplished without using the interchange operations of kind (1). If A is invertible it can be shown that, by permuting the rows of A in suitable fashion, the resulting matrix QA, Q being a permutation matrix, is expressible in the form QA = LU. Hence A = PLU where $P = Q^{-1}$. If A is singular an LU-decomposition cannot be achieved without also permuting the columns of A. Permuting the columns of A amounts to right multiplication by a permutation matrix and so in this case the best we can do is to write A = PLUQ.

Problems 1E

1. Prove Cramer's rule as stated in (1.6.1).

(*Hint*-show that the i-component of (adj A)b is $\sum_{j=1}^{n} C_{ji}b_j$ where C_{ji} denotes the cofactor corresponding to a_{ji} and use the properties of determinants to show that this sum also equals det A_i.)

2. Find the unique value of k for which the following system of linear

equations does not have a unique solution and determine the solution set in this case;

$$2x + y + z = 3$$

$$3x - y - z = 7$$

$$6x + y + kz = 11$$

3. Use the method described in *(1.6.2)* to solve the following;

$$\begin{pmatrix} 3 & -2 & -2 \\ 0 & 6 & -3 \\ 0 & 0 & 1 \end{pmatrix}\begin{pmatrix} 1 & 0 & 0 \\ 4 & 1 & 0 \\ -1 & 6 & 2 \end{pmatrix}\begin{pmatrix} x \\ y \\ z \end{pmatrix} = \begin{pmatrix} 2 \\ 0 \\ 2 \end{pmatrix}$$

4. Obtain an LU-decomposition for $\begin{pmatrix} 1 & 2 & 4 \\ -1 & -3 & 3 \\ 4 & 9 & 14 \end{pmatrix}$.

5. Obtain a PLU-decomposition for $\begin{pmatrix} 0 & -6 & 4 \\ 2 & 1 & 2 \\ 1 & 4 & 1 \end{pmatrix}$.

1.7 Some places where matrices are found

(a) Matrices in linear algebra

Matrices are a vital and essential part of the area of mathematics called *linear algebra* as we shall see fully in the next chapter. For the moment let us just observe that a mapping $f : \mathbb{R}^2 \longrightarrow \mathbb{R}^2$ given by $f(x,y) = (ax + by, cx + dy)$, where $\mathbb{R}^2$ denotes the set of all ordered pairs of real numbers and a,b,c, and d are constants, may be written succinctly in matrix form as $f(v) = Av$ where $v = \begin{pmatrix} x \\ y \end{pmatrix}$ and $A = \begin{pmatrix} a & b \\ c & d \end{pmatrix}$.

(b) The Jacobian matrix in calculus of several variables

Let $x_1, x_2, \ldots, x_r$ be a set of real variables and let $u_1, u_2, \ldots, u_r$ be a new set of variables. The $r \times r$ matrix with entries the partial derivatives $\partial x_i/\partial u_j$ is called the *Jacobian* matrix of change of variables.

This matrix is important in the differential calculus of several variables and its determinant is important in integration theory of functions of several variables.

(c) The Hessian matrix in calculus of several variables

Let $f(x_1,x_2,\ldots,x_n)$ be a function of n real variables. The *Hessian* matrix of f is the matrix which has entry $\partial^2 f/\partial x_i \partial x_j$, the second order partial derivative, in the (i,j)-place. This matrix is used in the analysis of the critical points of f.

(d) The Wronskian in differential equations

Consider the differential equation of order n

$$f^{(n)} + a_{n-1}f^{(n-1)} + \ldots\ldots + a_1 f^{(1)} + a_0 f = 0$$

where a_i, $i = 0,1,\ldots,n-1$ are constants and $f^{(i)}$ denotes the i-th derivative of the function f.

Let $f_1,f_2,\ldots,f_n$ be a set of functions each of which satisfies the above differential equation.

The determinant of the n×n matrix which has entry $f_j^{(i-1)}$ in the (i,j)-place for each $j = 1, 2,\ldots,n$, $i = 0, 1,\ldots, n-1$ is called the *Wronskian* of the set and provided it is non-zero the functions $f_1,f_2,\ldots f_n$ form an independent set of solutions to the differential equation.

(e) Matrices in control theory

Engineers of various kinds (electrical, mechanical, chemical etc.) deal with feedback control systems. A large class of such systems can be described mathematically by the following equations involving matrices;

$$\dot{x} = Ax + Bu, \; y = Cx$$

Here $x = (x_i)$, $u = (u_i)$, $y = (y_i)$ are column vectors of length n, m, p respectively, and A,B,C are matrices of size $n \times n$, $n \times m$, $p \times n$ respectively. These three vectors are each functions of time and $\dot{x}$ denotes the column vector of derivatives of x.

The vectors x,u,y are known as the *state vector,* the *input vector,* and the *output vector* respectively. The matrices A, B, C are known as the *system matrix,* the *input distribution matrix,* and the *measurement matrix* respectively. These three matrices are assumed to be independent of time.

(f) Matrices in the theory of graphs and networks

The abstract mathematical notion of a graph has applications in many practical situations, (e.g. electrical circuits, communications networks, transportation problems in management science, molecular structure in organic chemistry etc).

A *graph* consists of a finite set of vertices (also called *nodes)* and a finite set of edges (also called *branches*) such that each edge consists of a distinct pair of distinct vertices.

A *directed graph* is a graph such that each edge consists of an ordered pair of distinct vertices, i.e. each edge has a preferred direction.

(The graphs modelling electrical circuits have the connection points as vertices and pieces of the circuit consisting of an e.m.f source and a resistance as the edges. The graphs modelling chemical molecules have atoms as vertices and chemical bonds as the edges. The graphs modelling transportation problems could have different cities as vertices and routes between these cities as the edges.)

To any graph there is associated a matrix known as its *adjacency matrix* defined as follows;

We label the vertices V_1,V_2 etc. and the edges E_1,E_2 etc.

If the graph has n vertices then its adjacency matrix is the $n \times n$ symmetric matrix with entry 1 in the (i,j)-place whenever vertex V_i and vertex V_j form an edge and entry zero in all other places.

To any directed graph there is another matrix known as its *incidence matrix* defined as follows;

If the graph has n vertices and p edges then its incidence matrix is the $n \times p$ matrix whose k-th column has -1 in the i-th place and +1 in the j-th place where edge E_k is the ordered pair of vertices (V_i,V_j). All other entries of the k-th column are zero.

The directed graph in Figure 1.1 below has adjacency matrix

$$\begin{pmatrix} 0 & 1 & 0 & 1 \\ 1 & 0 & 1 & 1 \\ 0 & 1 & 0 & 1 \\ 1 & 1 & 1 & 0 \end{pmatrix} \text{ and incidence matrix } \begin{pmatrix} 1 & -1 & 0 & 0 & 0 \\ -1 & 0 & 1 & -1 & 0 \\ 0 & 0 & 0 & 1 & 1 \\ 0 & 1 & -1 & 0 & -1 \end{pmatrix}$$

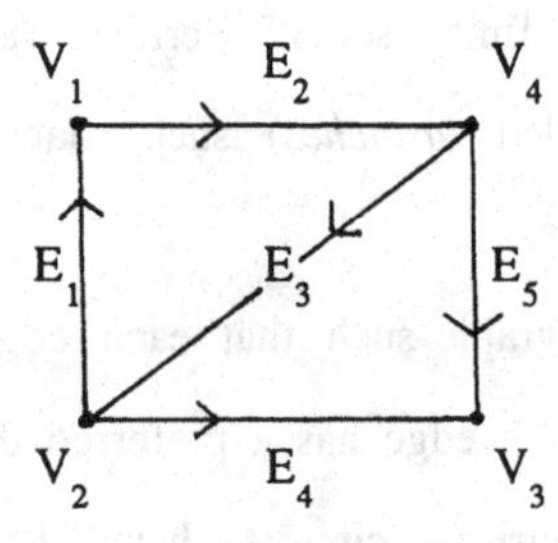

Figure 1.1

The adjacency matrix and the incidence matrix are both useful in the study of graphs and networks.

(g) Matrices in probability theory

The republic of Lowland has regular elections once a year and has three political parties labelled X,Y, and Z. The inhabitants of Lowland behave in a consistent fashion at elections and never vote for the same party at two successive elections. Those who vote for party X or party Y at one election are equally likely to vote next time for either of the two other parties. Those who vote for party Z at one election are twice as likely to vote next time for party X as to vote for party Y.

We construct a 3x3 matrix A as follows;

In the first column of A we place the probabilities that a person who voted for party X at one election will vote for parties X,Y,Z respectively at the next election. In the second (resp. third) column of A we place the probabilities that a person who voted for party Y (resp. Z) at one election will vote for parties X,Y,Z next time. The sum of the entries in each column of A equals one. The matrix A is called a *transition matrix*.

$$A = \begin{pmatrix} 0 & 1/2 & 2/3 \\ 1/2 & 0 & 1/3 \\ 1/2 & 1/2 & 0 \end{pmatrix}$$

Suppose that in the first ever election in Lowland each of the three parties gets the same number of votes, i.e. they each get one third of the total vote. The column vector $\begin{pmatrix} 1/3 \\ 1/3 \\ 1/3 \end{pmatrix}$ is called the *initial state vector*. The state of the parties after n elections will be described by the vector $A^n v$ where v is the initial state vector.

The above is an example of a *Markov process*. In chapter 5 we will see more about Markov processes.

(h) The covariance matrix in statistics

Let $X_1, X_2, \ldots, X_n$ be a set of random variables and let $E(X_i) = \mu_i$ for each i where E denotes expectation. Statisticians make use of the *covariance matrix* of the vector of random variables $(X_1, X_2, \ldots, X_n)$. This is the nxn matrix which has entry $E[(X_i - \mu_i)(X_j - \mu_j)]$ in the (i,j)-place.

APPENDIX TO CHAPTER 1

A proof of the properties of determinants

In this appendix we derive the properties of determinants listed in Proposition (1.4.2) and also prove Proposition (1.4.7). First we state and give the proof of each of the ten properties in (1.4.2).

Proposition 1.4.2

(1) $\det A = \sum_{j=1}^{n} (-1)^{j+1} a_{j1} M_{j1}$ where M_{ij} denotes the minor of the nxn matrix A corresponding to the entry a_{ij} of A. (See (1.4.1) for the definition of minor.)

i.e. det A can equivalently be obtained by expansion down the first column.

Proof

The proof is by induction on n.

It is easy to see that the result is true for n = 2 so let A be an nxn matrix and assume the result is true for all (n-1)x(n-1) matrices. Expansion of det A along the first row gives

$$\det A = \sum_{i=1}^{n} (-1)^{j+1} a_{1j} M_{1j}.$$

By the inductive assumption we can expand M_{1j} down the first column to get $M_{1j} = \sum_{i=2}^{n} (-1)^{i-1+1} a_{i1} d_{ij}$ where d_{ij} denotes the determinant of the (n-2)x(n-2) matrix obtained from A by deleting row 1, row i, column 1, and column j. Hence the only term in det A involving $a_{1j} a_{i1}$ is $(-1)^{i+j+1} a_{1j} a_{i1} d_{ij}$

Also by the inductive assumption $M_{i1} = \sum_{j=2}^{n} (-1)^{i+j+1} a_{1j} d_{ij}$. Thus, examining the expansion $\sum_{i=1}^{n} (-1)^{i+1} a_{i1} M_{i1}$, we see that the only term involving $a_{i1} a_{1j}$ is $(-1)^{i+j+1} a_{i1} a_{1j} d_{ij}$. This proves (1).

(2) det A = det A^t for all n×n matrices A.

Proof

Observe that the minor M_{ij} of A will equal the (j,i)-minor of the matrix A^t. Hence, after expanding along the first row, we see that det $A^t = \sum_{j=1}^{n} (-1)^{j+1} a_{j1} M_{j1}$ and by property (1) we see that det A is given by exactly the same expression. This proves (2).

(3) If A is upper or lower triangular then det A = $a_{11}a_{22}....a_{nn}$.

Proof

The proof is by induction on n.

The result is clearly true for n = 1 so let A be an n×n matrix and assume the result is true for all (n-1)×(n-1) matrices. If A is lower triangular expanding along the first row yields that det A = $a_{11}M_{11}$ = $a_{11}a_{22}......a_{nn}$ by the inductive assumption. If A is upper triangular expanding down the first column yields a similar result. This proves (3).

(4) If one row of A consists entirely of zeros then det A = 0.

Proof

The proof is by induction on n.

If n = 1 the result is clearly true so let A be an n×n matrix and assume the result is true for all (n-1)×(n-1) matrices A.

Now det A = $\sum_{j=1}^{n} (-1)^{j+1} a_{1j} M_{1j}$. If row 1 of A consists entirely of zeros then clearly det A = 0. If row i for some i > 1 consists entirely of zeros then each minor M_{1j} will be zero by the inductive assumption so that det A = 0. This proves (4).

(5) If B is the matrix obtained from A by multiplying one row of A by the scalar λ then detB = λ detA.

Proof

The proof is by induction on n.

The result is clearly true for n = 1 so let A be an n×n matrix and assume the result is true for all (n-1)×(n-1) matrices. If row 1 of A is multiplied by λ then each term in the expansion along the first row is multiplied by λ so that det B = λ det A.

If row i for some i > 1 is multiplied by λ then each minor appearing in the expansion along the first row will be multiplied by λ because of the inductive assumption. This gives det B = λ detA again and so (5) is proved.

(6) If B is the matrix obtained from A by interchanging two rows of A then det B = -det A.

Proof

Note first that the interchange of row i and row j can be achieved by successively interchanging an odd number of *adjacent* rows. (Specifically moving row i to below row j takes i - j interchanges of adjacent rows and then shifting row j into the original position of row i takes a further i - j - 1 interchanges.) Thus it suffices to prove the result in the case when two adjacent rows are interchanged.

Now suppose we interchange row i and row i + 1 of A. Write M_{ij} for the minors of A and R_{ij} for the minors of B. We observe that $M_{k1} = -R_{k1}$ for all $k \neq i, i+1$, that $a_{i1}M_{i1} = b_{(i+1)1}R_{(i+1)1}$, and that $b_{i1}R_{i1} = a_{(i+1)1}M_{(i+1)1}$.

Examining the expansions of det A and det B down the first column we see that det B = - det A. This proves (6).

(7) If two rows of A are identical then $\det A = 0$.

Proof

Interchanging the two identical rows yields $\det A = -\det A$ by property (6) and hence $\det A = 0$.

(8) If B is the matrix obtained from A by adding λ times one row of A to another row of A then $\det B = \det A$.

Proof

Let B be the $n \times n$ matrix having the same rows as A except for having λ times row i of A plus row j of A as its j-th row.

Let C be the $n \times n$ matrix having the same rows as A except for having λ times row i of A as its j-th row.

Observe first that $\det B = \det A + \det C$. (This can be seen by interchanging row j with row 1 for each of the three matrices A, B, C, finding $\det A$, $\det B$, $\det C$ by expansion along the first column, and using property (5).)

But $\det C = 0$ by properties (5) and (7) so that $\det B = \det A$.

(9) Properties (4), (5), (6), (7), and (8) remain valid if the word "row" is replaced everywhere by the word "column".

Proof

The columns of A are the rows of A^t and vice versa. The result now follows because of property (2).

(10) $\det AB = \det A \det B$ for any pair of $n \times n$ matrices A and B.

Proof

Assume first that A is non-singular. Then the argument in the proof of (1.3.5) shows that A is reducible to the identity matrix I by a sequence of elementary row operations. If we use only operations of type (1) and (3) we can reduce A to a diagonal matrix D with all of the diagonal entries of D being non-zero. Let r be the number of operations of type (1) used. (i.e. r is the number of row interchanges). Then, using properties (6) and (8), $\det A = (-1)^r \det D$.

Also $\det AB = (-1)^r \det DB$ since exactly the same sequence of elementary row operations transforms AB into DB.

Now let the diagonal entries of D be $\alpha_1, \alpha_2, \ldots, \alpha_n$. Multiplying B on the left by D amounts to multiplying row i of B by α_i for each $i = 1,2,\ldots,n$. Then by property (5) we see that $\det DB = \alpha_1\alpha_2 \ldots \alpha_n \det B = \det D \det B$. Thus $\det AB = (-1)^r \det D \det B = \det A \det B$.

Now suppose A is singular, i.e. A does not have an inverse. If also B is singular then the system of equations $Bx = 0$ has a non-zero solution. ($x = 0$ is always one solution of $Bx = 0$ and by (1.3.6) the solution of $Bx = 0$ is unique if and only if B is non-singular.) Hence the system $ABx = 0$ has a non-zero solution which implies that AB must be singular.

If B is non-singular then again AB must be singular for if AB is non-singular then (1.3.5) gives that $AB = E_1E_2 \ldots E_r$ for elementary matrices E_i, $i = 1,2,\ldots,r$ which implies that A is non-singular. ($A = E_1E_2 \ldots E_rB^{-1}$ which is non-singular by (1.3.3)(ii) and (1.3.4).)

The above argument shows that AB is singular whenever A is singular.

Now any singular matrix X must have zero determinant. (If X is singular then X is reducible to an echelon form E with a zero somewhere on the diagonal. Properties (5), (7), and (8) imply that $\det E = \alpha \det X$ for some non-zero scalar α, and det E = 0 by property (3).)

Thus whenever A is singular the equation det AB = det A det B is valid because each side of the equation is zero. This completes the proof of (10).

Proposition 1.4.7

Let A be an nxn matrix and let I be the nxn identity matrix. Then A (adj A) = (adj A) A = (det A) I.

Proof

Let $A = (a_{ij})$ have minors M_{ij} and cofactors $C_{ij} = (-1)^{i+j}M_{ij}$. Since adj A is the transpose of the matrix of cofactors of A the (i,j)-entry of the product A (adj A) will be $a_{i1}C_{j1} + a_{i2}C_{j2} + \ldots\ldots + a_{in}C_{jn}$.

For i = j = 1 the above sum will equal det A as it is precisely the definition of det A by expansion along the first row. Now suppose i > 1 and i = j. Let B be the matrix obtained from A by moving row i above row 1 and leaving all other rows fixed. B is obtained from A by successively interchanging row i with rows i-1, i-2, . . .,1. Hence $\det B = (-1)^{i-1}\det A$ using (6) of (1.4.2). Note that the (1,k)-minor of B will equal the minor M_{ik} of A. Expanding along the first row yields

$$\det B = a_{i1}M_{i1} - a_{i2}M_{i2} + \ldots \ldots + (-1)^{n-1}a_{in}M_{in}$$

Equating this with $\det B = (-1)^{i-1}\det A$ and using $C_{ij} = (-1)^{i+j}M_{ij}$ we see that $a_{i1}C_{i1} + a_{i2}C_{i2} + \ldots + a_{in}C_{in} = \det A$.

Now suppose $i \neq j$. Let G be the matrix obtained from A by moving row i above row 1, leaving the other rows fixed, and then replacing row j by a second copy of row i. Then det G = 0 by (7) of (1.4.2) as G has two identical rows. Note that the (1,k)-minor of G will equal $(-1)^{i-j-1}M_{jk}$. (The second copy of row i in G must be moved through i-j-1 rows to return to its rightful place.) Expanding along the first row yields

$$\det G = (-1)^{i-j-1}(a_{i1}M_{j1} - a_{i2}M_{j2} + \dots + (-1)^{n-1}a_{in}M_{jn}).$$

Using $\det G = 0$ and $C_{ij} = (-1)^{i+j}M_{ij}$ we see that

$$a_{i1}C_{j1} + a_{i2}C_{j2} + \dots\dots + a_{in}C_{jn} = 0.$$

This proves that A (adj A) = (det A) I.

Replacing A by A^t in this last equation, transposing the whole equation, and using (ii), (iii) of (1.1.11) and (2) of (1.4.2) yields that $(\text{adj } A^t)^t A = (\det A) I$. From the definition of adjugate and property (2) of (1.4.2) we see that adj $A^t = (\text{adj } A)^t$ for any matrix A.

Thus (adj A) A = (det A) I and the proof is complete.

Chapter 2

VECTOR SPACES AND LINEAR MAPS

In chapter 1 we introduced matrices, determinants and systems of linear equations. These are the technical tools needed in the subject of linear algebra. In this chapter we set up the conceptual framework necessary for the study of linear algebra. The fundamental notion is that of a vector space. The important functions to study are the linear maps between vector spaces. These are the maps which preserve the structure of vector spaces. Matrices can be viewed in a natural manner as linear maps and this viewpoint will be vital to our further study of matrices.

We begin the chapter with the definition of a vector space and give a large number of examples of vector spaces. We discuss the notion of linear independence of a set of elements of a vector space, the idea of a basis for a vector space, and prove the Basis Theorem. This leads on to the fundamental theorem which characterizes all finite-dimensional vector spaces. We carry on to introduce the notion of subspace of a vector space and then the more difficult but useful notion of quotient space. We then discuss in detail linear maps and their relationship with matrices. We define the rank and nullity of a linear map, prove the valuable Rank-Nullity Theorem, and give properties and equivalent characterizations of the rank of a matrix. Finally we consider vector spaces with an extra structure, namely that of an inner product. This extra structure allows us to utilize geometric properties such as orthogonality in the general context of vector spaces. At the very end of the chapter we make some miscellaneous comments concerning volumes in $\mathbb{R}^n$, the relationship between linear algebra and calculus, and the linearization of non-linear systems.

2.1 Vector spaces

Let F denote either the field of real numbers $\mathbb{R}$ or the field of complex numbers $\mathbb{C}$.

2.1.1 Definition

A *vector space over F* is a set V equipped with two operations called addition and scalar multiplication described as follows;

Given any two elements v_1 and v_2 of V there exists a unique element of V denoted $v_1 + v_2$ and called the sum of v_1 and v_2.

Given any element v of V and any element λ of F there exists a unique element of V denoted λv and called the scalar multiple of v by λ.

These two operations must satisfy the following eight axioms;

(i) $v_1 + v_2 = v_2 + v_1$ for all v_1,v_2 in V.

(ii) $(v_1 + v_2) + v_3 = v_1 + (v_2 + v_3)$ for all v_1,v_2,v_3 in V.

(iii) There exists a unique element of V called the zero vector and denoted O_V with the property that $v + O_V = O_V + v = v$ for all elements v in V.

(iv) For each element v of V there exists an element denoted - v and called the additive inverse of v with the property that

$v + (-v) = (-v) + v = O_V$.

(v) $\lambda_1(\lambda_2 v) = (\lambda_1\lambda_2)v$ for all λ_1,λ_2 in F,and all v in V.

(vi) $(\lambda_1 + \lambda_2)v = \lambda_1 v + \lambda_2 v$ for all λ_1,λ_2 in F, and all v in V.

(vii) $\lambda(v_1 + v_2) = \lambda v_1 + \lambda v_2$ for all λ in F, and all v_1,v_2 in V.

(viii) $1v = v$ for all elements v in V, where 1 denotes the unity element of the field F.

The elements of V are usually referred to as *vectors* while the elements of F are referred to as *scalars*.

In this book we are taking $F = \mathbb{R}$ or $F = \mathbb{C}$ only. Mathematicians do indeed consider vector spaces over other fields, (e.g. finite fields, the field of rational numbers etc.), the definition being exactly as above.

2.1.2 **Comment**

The definition given may seem rather daunting at first. However we will give a large number of examples which will illustrate the fact that vector space structures do occur frequently in mathematics.

2.1.2 **Examples of vector spaces**

Example 1

Let $V = \mathbb{R}^n = \{ (x_1,x_2,. . . . .,x_n) : \text{each } x_i \in \mathbb{R} \}$, the set of all the ordered n-tuples of real numbers.

Define addition on V by

$(x_1,x_2, . . .,x_n) + (y_1,y_2, . . .,y_n) = (x_1+y_1,x_2+y_2, . . .,x_n+y_n)$ for all $x_i,y_i \in \mathbb{R}$, $i = 1,2,. . .,n$.

Define scalar multiplication on V by

$\lambda(x_1,x_2, . . .,x_n) = (\lambda x_1,\lambda x_2, . . . ,\lambda x_n)$ for all $\lambda \in \mathbb{R}$, and all $x_i \in \mathbb{R}$, $i = 1,2, . . ,n$.

It is straightforward to check that the eight vector space axioms are satisfied so that $\mathbb{R}^n$ is a vector space over $\mathbb{R}$. The zero vector of $\mathbb{R}^n$ is the n-tuple $(0,0, ...,0)$ and the additive inverse of $(x_1,x_2,. . . ,x_n)$ is $(-x_1,-x_2, . . .,-x_n)$.

For $n = 1$, 2, or 3 we have a geometric intuitive picture of $\mathbb{R}^n$. (When $n = 1$ it is the real line, when $n = 2$ it is the plane, and when $n = 3$ it is ordinary three-dimensional space in which we live.) For larger values of n we cannot visualize $\mathbb{R}^n$ but by analogy with the situation for n = 1,2, or 3 we can have some useful geometric concepts in $\mathbb{R}^n$.

Similarly $V = \mathbb{C}^n$, the set of all ordered n-tuples of complex numbers, is a vector space over $\mathbb{C}$.

Example 2

Let V be the set of all polynomials in one variable x with coefficients from $\mathbb{R}$. A typical element of V is an expression of the form $\sum_{i=1}^{n} a_i x^i$ where each $x_i \in \mathbb{R}$. The integer n is called the *degree* of the polynomial and it may take any non-negative integer value. The real numbers a_i are called the *coefficients* of the polynomial. Two polynomials are said to be *equal* if and only if all their corresponding coefficients are equal.

Define addition in V by $\sum_{i=1}^{n} a_i x^i + \sum_{i=1}^{n} b_i x^i = \sum_{i=1}^{n} (a_i + b_i)x^i$ and scalar multiplication by $\lambda \left(\sum_{i=1}^{n} a_i x^i \right) = \sum_{i=1}^{n} \lambda a_i x^i$ for all $\lambda \in \mathbb{R}$.

It is a straghtforward exercise to verify that V satisfies the eight vector space axioms so that V is a vector space over $\mathbb{R}$. (The zero vector is the zero polynomial, i.e. the polynomial with a_i equal to zero for all i.)

Similarly, by allowing the coefficients of the polynomials to be complex numbers, we can obtain a vector space over $\mathbb{C}$.

Example 3

Let V be the set of all polynomials of degree at most 6. Defining addition and scalar multiplication as in the previous example we obtain a vector space. Note that the operations of addition and scalar multiplication on elements of V always yield elements of V.

Similarly for any fixed integer n we may have the vector space of all polynomials of degree at most n.

Example 4

Let A be an m×n matrix with real entries and let V be the solution set of the system of linear equations $Ax = 0$ where x denotes a column vector of unknowns in $\mathbb{R}^n$. Then V is a subset of $\mathbb{R}^n$ and the operations of addition and scalar multiplication on elements of V always yield elements of V. Thus V is a vector space over $\mathbb{R}$.

Example 5

Let V be the set of all continuous functions $f:\mathbb{R} \longrightarrow \mathbb{R}$. Define addition on V by $(f + g)(x) = f(x) + g(x)$ for all $x \in \mathbb{R}$ and scalar multiplication on V by $(\lambda f)(x) = \lambda f(x)$ for all $x \in \mathbb{R}$. (Note that $f + g$ and λf are continuous by the properties of continuous functions). Then V is a vector space over $\mathbb{R}$, the zero vector being the function with constant value zero.

Example 6

Let V be the set of all solutions of the differential equation

$$\frac{d^2y}{dx^2} + 4\frac{dy}{dx} - 3y = 0$$

The sum of two solutions of this differential equation is again a solution and the scalar multiple of a solution is again a solution. Then V is a vector space over $\mathbb{R}$ whose zero vector is the solution $y = 0$.

Example 7

Let V be the set of all m×n matrices with entries from $\mathbb{R}$. The operations of addition and scalar multiplication defined in (1.1.1) make V into a vector space over $\mathbb{R}$. The zero vector is the matrix with each entry equal to zero.

Similarly if we allow the entries of the matrices to be complex numbers we obtain a vector space over $\mathbb{C}$.

Example 8

Let V be the set of all upper triangular $n \times n$ matrices with entries in F where either $F = \mathbb{R}$ or $F = \mathbb{C}$. Then addition and scalar multiplication of elements of V again yields elements of V so that V is a vector space over F.

Example 9

Let V be the set of all symmetric $n \times n$ matrices with entries in F where either $F = \mathbb{R}$ or $F = \mathbb{C}$. Then V is a vector space over F since addition and scalar multiplication of elements of V again yields elements of V.

Example 10

The zero vector space is the vector space containing only one element, this one element necessarily being the zero vector. We will write $V = 0$ to mean that V is the zero vector space.

We also use the symbol 0 for the zero element of the field F but this should not cause any difficulties since it should be clear from the context in which we are working as to which of the two notions the symbol 0 stands for.

2.1.3 Consequences of the vector space axioms

(a) Cancelling on the left and on the right is allowed in a vector space, i.e. $v + w_1 = v + w_2$ implies $w_1 = w_2$ for v, w_1, w_2 in V, and $w_1 + v = w_2 + v$ implies $w_1 = w_2$ for v, w_1, w_2 in V.

(b) $0v = 0_V$ for all $v \in V$, 0 being the zero element of F.

(c) $\lambda 0_V = 0_V$ for all $\lambda \in F$.

(d) $(-1)v = -v$ for all $v \in V$.

(e) $\lambda v = 0_V$ implies $\lambda = 0$ or $v = 0_V$ where $\lambda \in F$, $v \in V$.

Proof

(a) If $v + w_1 = v + w_2$ then $(-v) + (v + w_1) = (-v) + (v + w_2)$.

Axioms (ii) and (iv) then yield that $0_V + w_1 = 0_V + w_2$ and axiom (iii) now gives $w_1 = w_2$. Cancellation on the right is proved in a similar way.

(b) Since $0 + 0 = 0$ in F we see that $(0 + 0)v = 0v$ for any $v \in V$.

Axiom (vi) shows that $0v + 0v = 0v$ and axiom (iii) yields that $0v + 0v = 0v + 0_V$. The desired result $0v = 0_V$ now follows by using (a) and cancelling $0v$ on each side of the equation.

The proofs of (c), (d), and (e) are left as an exercise for the reader to be done as problem 2 of problems 2A below.

Problems 2A

1. Determine whether or not each of the following is a vector space:

(i) The set of all differentiable functions $f:\mathbb{R} \longrightarrow \mathbb{R}$ with addition and scalar multiplication of functions as in Example 5.

(ii) The set $\mathbb{R}^2$ with the usual addition but with scalar multiplication defined by $\lambda(x,y) = (\lambda y, \lambda x)$ for $\lambda \in \mathbb{R}$.

(iii) The set of all $n \times n$ matrices of trace zero with the usual operations of addition and scalar multiplication.

(iv) The set of all $n \times n$ matrices of determinant zero with the usual operations of addition and scalar multiplication.

2. Deduce (c), (d),and (e) of (2.1.3) from the vector space axioms.

[*Hint-* For (c) write $0_V + 0_V = 0_V$ and use axioms (iii) and (vii).

For (d) write $(-1 + 1)v = 0v = 0_V$. For (e) assume $\lambda \neq 0$ and multiply the equation $\lambda v = 0_V$ by λ^{-1}.]

2.2 Bases and dimension

2.2.1 Definition

Let $v_1, v_2, \ldots, v_k$ be a set of vectors in a vector space over F. Any expression of the form $\alpha_1 v_1 + \alpha_2 v_2 + \ldots + \alpha_k v_k$, where each $\alpha_i \in F$, is called *a linear combination* of the vectors $v_1, v_2, \ldots, v_k$.

2.2.2 Definition

A set of vectors $\{v_1, v_2, \ldots, v_k\}$ in a vector space V is said to be *linearly independent* provided that the zero vector of V is not expressible as a non-trivial linear combination of the vectors $v_1, v_2, \ldots, v_k$. i.e. if $\alpha_1 v_1 + \alpha_2 v_2 + \ldots + \alpha_k v_k = 0_V$ for scalars $\alpha_i \in F$ then each $\alpha_i = 0$.

2.2.3 Definition

A set of vectors $\{v_1, v_2, \ldots, v_k\}$ in a vector space V is said to be *linearly dependent* provided that the set is not linearly independent.

2.2.4 Remark

Any set of vectors which includes the zero vector must be a linearly dependent set because $\lambda 0_V = 0_V$ for all $\lambda \in F$.

2.2.5 Definition

A set of vectors $\{v_1, v_2, \ldots, v_k\}$ in a vector space V is said to be a *spanning set* for V provided that each vector in V is expressible as a linear combination of $v_1, v_2, \ldots, v_k$.

We say that such a set *spans* V.

2.2.6 Definition

A set of vectors $\{v_1, v_2, \ldots, v_k\}$ in a vector space V is said to be a *basis* for V provided that it is both a linearly independent set and a spanning set for V.

(Note that the plural of basis is bases.)

2.2.7 Example

Let $V = F^n$ where $F = \mathbb{R}$ or $\mathbb{C}$ and let $e_i \in V$ be the n-tuple which has 1 in the i-th place and zero elsewhere. Then $\{e_1,e_2,...,e_n\}$ is a basis for V as a vector space over F. It is usually known as the *standard basis* for F^n.

2.2.8 Definition

A vector space V is said to be *finite-dimensional* provided that there exists a finite set of vectors in V which is a basis for V.

Otherwise V is said to be *infinite-dimensional.*

Example 2 of (2.2.1) is an infinite-dimensional vector space. See problem 5 of problems 2B for a proof.

Our main interest in this book will be in finite-dimensional vector spaces.

2.2.9 The replacement lemma

Let $\{w_1,w_2,...,w_n\}$ be a basis of the finite-dimensional vector space V over F. Let v be any non-zero vector in V. Then there exists w_i such that if we replace w_i by v then we still have a basis of V.

Proof

Since $\{w_1,w_2,...,w_n\}$ spans V we may write $v = \sum_{i=1}^{n} \alpha_i w_i$ for scalars α_i, $i = 1,2,...,n$. Now at least one α_i is non-zero because v is a non-zero vector.

Assume $\alpha_1 \neq 0$. Then $w_1 = (1/\alpha_1)v - (1/\alpha_1)\sum_{i=2}^{n} \alpha_i w_i$.

It follows that $\{v,w_2,w_3,...,w_n\}$ spans V.

To show that $\{v,w_2,w_3,...,w_n\}$ is linearly independent let

$\beta_1 v + \beta_2 w_2 + \beta_3 w_3 + ... + \beta_n w_n = 0_V$ for scalars $\beta_1,\beta_2,...,\beta_n$.

Then $\beta_1\alpha_1 w_1 + \sum_{i=2}^{n} (\beta_1\alpha_i + \beta_i)w_i = 0_V$ and so using the linear independence of $\{w_1,w_2,...,w_n\}$ we see that $\beta_1\alpha_1 = 0$ and that $\beta_1\alpha_i + \beta_i = 0$ for each

$i = 2,3,\ldots,n$. Since $\alpha_1 \neq 0$ we get that each $\beta_i = 0$ for $i = 1,2,3,\ldots,n$ and thus $\{v,w_2,w_3,\ldots,w_n\}$ is linearly independent.

If $\alpha_1 = 0$ then a similar argument will work using an α_i which is non-zero. This completes the proof.

2.2.10 **The Basis Theorem**

Let V be a finite-dimensional vector space over F. All bases of V have the same number of elements.

Proof

If $\{v_1,v_2,\ldots,v_m\}$ and $\{w_1,w_2,\ldots,w_n\}$ are two bases of V we must show that $m = n$.

We can assume without any loss of generality that $m \leq n$. Applying the replacement lemma (2.2.9) with $v = v_1$ we may replace one of the vectors w_i by v_1 and still have a basis of V. Applying (2.2.9) again to this new basis and with $v = v_2$ we may replace one of the vectors w_i by v_2 and still have a basis.

(Note that we can definitely choose one of the w_i rather than v_1 to be replaced by v_2. This is because v_2 cannot be expressible in terms of v_1 alone since the vectors v_i form a linearly independent set).

Proceeding like this we can successively insert all of the vectors v_i in place of m of the vectors w_i. (As before, linear independence of the vectors v_i ensures that we can always select one of the vectors w_i to be replaced.)

If $m < n$ then the resulting basis contains all the vectors v_i and at least one of the vectors w_i. Any vector w_i is a linear combination of $\{v_1,v_2,\ldots,v_m\}$ since this set spans V. The elements of a basis are linearly independent and so this leads to a contradiction unless $m = n$. This completes the proof.

2.2.11 **Definition**

The *dimension* of a finite-dimensional vector space V is the number of elements in a basis of V.

This definition is meaningful because of (2.2.10).

We will write dim V as short for the dimension of V.

2.2.12 **Examples**

(i) F^n has dimension n as we have seen in (2.2.7) that the standard basis of F^n has n elements.

(ii) The vector space of all m×n matrices which we met as Example 7 of (2.1.2) has dimension mn.

A basis for this vector space is the set of matrices of the form E_{ij} where E_{ij} has entry 1 in the (i,j)-place and zero elsewhere.

(iii) The vector space of all polynomials of degree at most 6 which we met as Example 3 of (2.1.2) has dimension 7. A basis for this space is $\{1, x, x^2, x^3, x^4, x^5, x^6\}$.

2.2.13 **Proposition**

The following properties hold in a vector space V of dimension n ;

(a) Any subset of V which contains more than n elements must be linearly dependent.

(b) Any linearly independent subset of V may be extended to a basis of V.

i.e. if $\{u_1, u_2, \ldots, u_k\}$ is a linearly independent set in V and $1 \leq k \leq n$ then there exist vectors $w_1, w_2, \ldots, w_{n-k}$ in V such that the set $\{u_1, u_2, \ldots, u_k, w_1, w_2, \ldots, w_{n-k}\}$ is a basis of V.

(c) Let $\{v_1, v_2, \ldots, v_n\}$ be a basis of V. Then the expression for any vector $v \in V$ as a linear combination of the vectors in this basis is unique.

i.e. if $v = \sum_{i=1}^{n} \alpha_i v_i$ for scalars α_i and also $v = \sum_{i=1}^{n} \beta_i v_i$ for scalars β_i then $\alpha_i = \beta_i$ for each $i = 1,2,\ldots,n$.

Proof

(a) Let S be a subset of V containing more than n elements. Suppose that S is linearly independent. Then successive use of the replacement lemma (2.2.9) enables us to replace any basis of V by the first n elements of S and still have a basis. This contradicts the linear independence of S if S has more than n elements. Hence S must be linearly dependent.

(b) Take any basis of V and by successive use of the replacement lemma (2.2.9) we may replace k elements of the basis by $u_1, u_2, \ldots, u_k$.

As in the proof of (2.2.10) the linear independence of $\{u_1, \ldots, u_k\}$ ensures that it is always members of the original basis that can be deleted rather than previously inserted members of $\{u_1, u_2, \ldots, u_k\}$.

(c) If $v = \sum_{i=1}^{n} \alpha_i v_i = \sum_{i=1}^{n} \beta_i v_i$ then $\sum_{i=1}^{n} (\alpha_i - \beta_i) v_i = 0$. The linear independence of $\{v_1, v_2, \ldots, v_n\}$ implies that $\alpha_i - \beta_i = 0$ for each $i = 1,2,\ldots,n$. Thus the expression of the vector v as a linear combination of the basis vectors is unique.

2.2.14 **Comment**

If we know that our vector space V has dimension n then any set of n vectors in V which is linearly independent will form a basis of V. This follows from (b) of (2.2.13).

If $V = F^n$ then to prove that a given set of n vectors in F^n forms a basis of F^n it suffices to show that the matrix B with these vectors as its rows has determinant non-zero. (If the rows of B are linearly dependent then some linear combination of the rows will equal the zero vector so that by properties (8) and (4) of determinants in (1.4.2) we see that det B = 0.)

Problems 2B

1. Determine whether or not each of the following sets of vectors is linearly independent.

(i) $\{(1,0,2,1),(1,3,2,1),(4,1,2,2)\}$ in $\mathbb{R}^4$.

(ii) $\{(1,2,6),(-1,3,4),(-1,-4,-2)\}$ in $\mathbb{R}^3$.

(iii) $\{u + v, v + w, w + u\}$ in a space V given that $\{u,v,w\}$ is linearly independent in V.

2. Show that the vector space of all upper triangular $n \times n$ matrices with real entries has dimension $n(n + 1)/2$ as a vector space over $\mathbb{R}$.

3. Show that the vector space of all symmetric $n \times n$ matrices with real entries has dimension $n(n + 1)/2$ as a vector space over $\mathbb{R}$.

4. Show that the vector space of all polynomials with real coefficients and of degree at most n has dimension $n + 1$ as a vector space over $\mathbb{R}$.

5. Show that the vector space of all polynomials as in Example 2 of (2.1.2) is infinite-dimensional. [*Hint* - show that $\{1, x, x^2,.., x^n\}$ is linearly independent for any value of n.]

6. Determine whether or not each of the following sets is a basis.

(i) $\{(5,3,7), (1,-3,6), (0,3,1)\}$ for $\mathbb{R}^3$.

(ii) $\{(1,-1,1,-1), (2,2,2,2), (1,2,1,2), (1,2,3,4)\}$ for $\mathbb{R}^4$.

(iii) $\{1, 1 + x, 1 + x^2, 1 + x^3, 1 + x^4, 1 + x^5, 1 + x^6\}$ for the vector space of all polynomials of degree at most 6.

2.3 Linear maps and Isomorphisms

We begin this section by recalling some of the standard terminology on functions.

2.3.1 Injective, surjective, and bijective functions

Let $f:A \longrightarrow B$ be a function from the set A to the set B, i.e. to each $x \in A$ we associate an element of B which we denote by $f(x)$. Functions will often be referred to as *mappings* or *maps*.

The set A is called the *domain* of f, the set B is called the *codomain* of f.

The *image* of f is $\{f(x) : x \in A\}$ which is a subset of B. i.e. the image of f is the set of values taken by f.

The function $f:A \longrightarrow B$ is said to be *injective* (or *one-one*) provided that $f(x_1) = f(x_2)$ implies $x_1 = x_2$, i.e. f does not send two different elements of A to the same element of B.

The function $f:A \longrightarrow B$ is said to be *surjective* (or *onto*) provided that given any element $y \in B$ there exists an element $x \in A$ such that $y = f(x)$, i.e. the image of f is all of B.

The function $f:A \longrightarrow B$ is said to be *bijective* provided that f is both injective and surjective. Note that a bijective function gives a one-one correspondence between the domain A and the codomain B.

Recall that the function $f:A \longrightarrow B$ is said to *invertible* provided that there exists a function $g:B \longrightarrow A$ such that the composite functions $g \circ f$ and $f \circ g$ are the identity functions on A and B respectively. The functions f and g are then said to be *inverses* of each other.

2.3.2 **Exercise**

Show that $f:A \longrightarrow B$ is invertible if and only if f is bijective.

2.3.3 **Definition**

The function $f:V \longrightarrow W$, where V and W are vector spaces over F, is said to be *a linear map* provided that

(i) $f(v_1 + v_2) = f(v_1) + f(v_2)$ for all v_1, v_2 in V.

(ii) $f(\lambda v) = \lambda f(v)$ for all v in V and all λ in F.

Linear maps are sometimes referred to as *linear transformations*.

2.3.4 **Remark**

Properties (i) and (ii) of a linear map imply that if $\sum_{i=1}^{n} \alpha_i v_i$ is some linear combination of vectors $v_1, v_2, .., v_n$ in V then $f(\sum_{i=1}^{n} \alpha_i v_i) = \sum_{i=1}^{n} \alpha_i f(v_i)$.

In particular the values which the linear map f takes on some basis of V will determine f completely.

The above property gives a useful method of constructing linear maps when the domain V is finite-dimensional. We choose some basis $\{v_1, v_2, . . , v_n\}$ of V and specify the values $f(v_i)$ for $i = 1, 2, . . , n$. Then since any element $v \in V$ is uniquely expressible as $v = \sum_{i=1}^{n} \alpha_i v_i$ we define f by $f(v) = \sum_{i=1}^{n} \alpha_i f(v_i)$. This process is often known as *extending by linearity*.

We should also point out the fact that $f(0_V) = 0_W$ for any linear map $f:V \longrightarrow W$. This follows by putting $\lambda = 0$ in (ii) and using (b) of (2.1.3).

2.3.5 **Examples of linear maps**

(i) The map $f:\mathbb{R}^2 \longrightarrow \mathbb{R}^2$, $f(x,y) = (ax + by, cx + dy)$, which we met in (1.7) is a linear map and can be written more succinctly as $f(v) = Av$ where $A = \begin{pmatrix} a & b \\ c & d \end{pmatrix}$, $v = \begin{pmatrix} x \\ y \end{pmatrix}$.

(ii) Let A be any m×n matrix with real entries and define a map $f:\mathbb{R}^n \longrightarrow \mathbb{R}^m$ by $f(v) = Av$ for each column vector $v \in \mathbb{R}^n$. The properties of matrix multiplication ensure that f is a linear map.

(iii) Let V be the vector space of all polynomials as defined in Example 2 of (2.1.2). Define a map $f:V \longrightarrow V$ by $f(p(x)) = p'(x)$, the derivative of the polynomial p(x). The properties of differentiation ensure that f is linear.

(iv) Let V be the space of all n×n matrices with entries in F and define a map $f:V \longrightarrow V$ by $f(A) = \text{trace } A$. The properties of trace, see (1.1.13), ensure that f is linear.

2.3.6 **Definition**

The linear map $f:V \longrightarrow W$ is said to be an *isomorphism* provided that it is bijective.

The vector spaces V and W are said to be *isomorphic* if and only if there exists an isomorphism $f:V \longrightarrow W$.

(It is an easy exercise to check that the inverse $g : W \longrightarrow V$ of an isomorphism $f:V \longrightarrow W$ is linear and so is also an isomorphism.)

We will use the notation $V \cong W$ as shorthand for saying that the vector spaces V and W are isomorphic.

2.3.7 **Theorem**

Let V be a vector space over F and suppose that V has dimension n. Then V is isomorphic to F^n.

Proof

Choose a basis $\{v_1,v_2,. . .,v_n\}$ of V and let $\{e_1,e_2,. . .,e_n\}$ be the standard basis of F^n as defined in (2.2.7). We will define a linear map $f:V \longrightarrow \mathbb{R}^n$ by requiring $f(v_i) = e_i$ for each i = 1, 2, . . , n and then extending by linearity as in (2.3.4). We can describe f explicitly as follows ;

Let $v \in V$ be written as $v = \sum_{i=1}^{n} \alpha_i v_i$ where $\alpha_1,\alpha_2,. . ,\alpha_n$ are uniquely determined by v. Then $f(v) = (\alpha_1,\alpha_2,. . .,\alpha_n)$.

Note that f is injective since $f\left(\sum_{i=1}^{n} \alpha_i v_i\right) = f\left(\sum_{i=1}^{n} \beta_i v_i\right)$ implies that $\alpha_i = \beta_i$ for each i = 1,2,. . ,n.

Also f is surjective since any element $w = (\alpha_1,\alpha_2,. . ,\alpha_n) \in F^n$ may be written as $w = f\left(\sum_{i=1}^{n} \alpha_i v_i\right)$.

The linear map f is thus an isomorphism. Note that the inverse of f is the map $g:F^n \longrightarrow V$ given by defining $g(e_i) = v_i$ for each i = 1,2,. . ,n and then extending by linearity.

2.3.8 **Comment**

The isomorphism in (2.3.7) is not canonical insofar as that it depends on the choice of basis for V. A different choice of basis will yield a different isomorphism from V to F^n.

2.3.9 **Comment**

As a result of (2.3.7) any n-dimensional vector space over $\mathbb{R}$ can be thought of as being like a copy of $\mathbb{R}^n$. Thus in spite of the abstract definition of vector space given in (2.1) we can think of finite-dimensional vector spaces in a more concrete manner.

2.4 Subspaces and quotient spaces

2.4.1 Definition

A subset U of a vector space V over F is said to be a *subspace* of V provided that U is itself a vector space .

Note that a subset U of V will be a subspace of V if and only if U is closed under the two operations of addition and scalar multiplication in V. The properties in the eight vector space axioms of (2.1) are then automatically inherited by U since they hold in V. Thus U is a subspace of V if and only if the following two conditions hold ;

(i) $u_1 + u_2 \in U$ whenever $u_1, u_2 \in U$.

(ii) $\lambda u \in U$ whenever $\lambda \in F$, $u \in U$.

2.4.2 Proposition

Let U be a subspace of the n-dimensional vector space V.

Then dim U $\leq$ dim V and dim U = dim V if and only if U = V.

Proof

By (2.2.13)(b) any basis of U can be extended to a basis of V. Hence dim U $\leq$ dim V. Clearly U = V if and only if any basis of U is also a basis of V.

2.4.3 Examples of subspaces

Example 1

Let $V = \mathbb{R}^2$ and $U = \{(x,7x) \; ; \; x \in \mathbb{R}\}$. Then U is a subspace of dimension one. Geometrically U is the line $y = 7x$, i.e. the line of slope 7 which passes through the origin. Similarly for any fixed $\lambda \in \mathbb{R}$ the line $y = \lambda x$ is a subspace of $\mathbb{R}^2$ of dimension one.

Example 2

Let $V = \mathbb{R}^3$ and $U = \{ (x,y,z); \; x + y + z = 0\}$. Then U is a subspace of dimension 2. A basis for U is $\{(1,0,-1),(0,1,-1)\}$ since this set is clearly linearly independent and it spans U as any element of U is of the form $(x,y,-x-y) = x(1,0,-1) + y(0,1,-1)$. Geometrically U is a plane passing through the origin of $\mathbb{R}^3$. Similarly any plane through the origin in $\mathbb{R}^3$ will be a subspace of dimension 2.

In a similar fashion any line through the origin of $\mathbb{R}^3$ will be a subspace of dimension 1, the points on the line being the set of all scalar multiples of some fixed vector in $\mathbb{R}^3$.

Examples 1 and 2 illustrate the fact that a vector space of dimension n will have infinitely many subspaces of dimension k for every k, $1 \leq k < n$.

Example 3

The vector space in example 4 of (2.1.2), i.e. the solution set of the homogeneous system of linear equations $Ax = 0$ where A is an $m \times n$ matrix, is a subspace of $\mathbb{R}^n$. Its dimension as a vector space will be precisely the dimension of a solution set as defined in (1.2.10).

Example 4

The vector space of all polynomials of degree at most 6 is a subspace of the vector space V of all polynomials. (See Examples 2 and 3 of (2.1.2)). This subspace has dimension 7 by (iii) of (2.1.12) while V is infinite-dimensional by problem 5 of Problems 2B.

Example 5

The vector space of all symmetric $n \times n$ matrices and the vector space of all upper triangular $n \times n$ matrices, (see Examples 9 and 8 of (2.1.2)), are each subspaces of the vector space V of all $n \times n$ matrices, (see Example 7 of (2.1.2)). These two subspaces each have dimension $n(n + 1)/2$ by problems 2 and 3 of Problems 2B while V has dimension n^2 by (2.2.12) (ii).

2.4.4 **Proper subspaces**

Let V be any vector space. Then V may be viewed as a subspace of itself. Also the subset $\{O_V\}$ is a subspace of V called the zero subspace and denoted 0.

A *proper subspace* of V is a subspace U of V such that $U \neq V$ and $U \neq 0$.

2.4.5 **Subspace spanned by a set of vectors**

Let S be any set of vectors in a vetor space V. *The subspace spanned by S* is the set of all linear combinations of vectors in S. It is the smallest subspace of V which contains S.

Let U and W each be subspaces of a vector space V. Then we write U + W for the subspace spanned by the union of the sets U and W. It is the smallest subspace of V which contains both U and W.

In fact $U + W = \{ x + y : x \in U, y \in W \}$. See problem 4 in Problems 2C for information about the dimension of U + W.

2.4.6 **Quotient spaces**

Let U be a subspace of the vector space V and let $v \in V$. The subset $\{ v + u : u \in U \}$ is called the *coset* of U in V determined by v and is denoted by v + U. Given two elements v_1, v_2 of V it can be shown that the cosets $v_1 + U$ and $v_2 + U$ either coincide or else have no elements in common. (Suppose that $v_1 + U$ and $v_2 + U$ have an element in common. It follows easily that $v_1 - v_2 \in U$ and from this that $v_1 + U = v_2 + U$.)

We write V/U for the set of all distinct cosets v + U.

We define an operation of addition on V/U by

$$(v_1 + U) + (v_2 + U) = (v_1 + v_2) + U \text{ for } v_1, v_2 \in V$$

We define an operation of scalar multiplication on V/U by

$$\lambda(v + U) = \lambda v + U \text{ for } \lambda \in F, v \in V.$$

It is straightforward to check that these two operations are well-defined, i.e. independent of the particular choice of element v representing the coset v + U, and that they make V/U into a vector space over F. (The zero vector in V/U is the coset O_V + U, usually written just as U and the additive inverse of v + U is (-v) + U.)

The space V/U is called the *quotient space* of V by U.

2.4.7 Proposition

If V is finite-dimensional then so is the quotient space V/U and

$$\dim V/U = \dim V - \dim U.$$

Proof

See problem 7 of Problems 2C.

Problems 2C

1. Let U = { $A \in M_nF$: trace A = 0 }. Show that U is a subspace of M_nF and its dimension is $n^2 - 1$.

2. Let U_1 and U_2 be subspaces of a vector space V. We say that V is the *direct sum* of U_1 and U_2, notation $V = U_1 \oplus U_2$, provided that each element of V has a unique expression in the form v = x + y where $x \in U_1$ and $y \in U_2$.

Show that $V = U_1 \oplus U_2$ if and only if $U_1 \cap U_2 = 0$ and each element of V is expressible in the form v = x + y where $x \in U_1$ and $y \in U_2$.

Show that if $V = U_1 \oplus U_2$ is finite-dimensional then

$$\dim V = \dim U_1 + \dim U_2.$$

3. Show that $U_1 = \{A \in M_nF : A^t = A\}$ and $U_2 = \{A \in M_nF : A^t = -A\}$ are each subspaces of M_nF.

Show that V is the direct sum of U_1 and U_2. [*Hint* - if $A \in M_nF$ then $A = (1/2)(A + A^t) + (1/2)(A - A^t)$]

Determine the dimensions of U_1 and U_2.

4. Let U and W be subspaces of the vector space V. Let $U + W$ be the subspace defined in (2.4.5) and let $U \cap W = \{ v \in V : v \in U \text{ and } v \in W \}$.

Show that $U \cap W$ is a subspace and that if V is finite-dimensional then

$$\dim (U + W) = \dim U + \dim W - \dim U \cap W.$$

[*Hint*-take a basis B for $U \cap W$, extend it in two different ways, firstly to a basis B_1 of U and secondly to a basis B_2 of W, and show that $B_1 \cup B_2$ is a basis for $U + W$. The result follows from the elementary set theoretic fact that $|B_1 \cup B_2| = |B_1| + |B_2| - |B_1 \cap B_2|$.]

5. Let U be a subspace of the vector space V. Let $v_1 + U$ and $v_2 + U$ be cosets as defined in (2.4.6).

Show that if $v_1 - v_2 \in U$ then $v_1 + U = v_2 + U$ and that if $v_1 - v_2 \notin U$ then $(v_1 + U) \cap (v_2 + U)$ is the empty set.

Show that $\dim(V/U) = \dim V - \dim U$.

[*Hint*-for the last part take a basis $\{u_1, u_2, \ldots, u_k\}$ of U and extend it to a basis $\{u_1, u_2, \ldots, u_k, v_1, v_2, \ldots, v_r\}$ of V Show that $\{v_1 + U, v_2 + U, \ldots, v_r + U\}$ is a basis of V/U.]

6. Let U be the subspace of $\mathbb{R}^4$ spanned by (1,0,3,2), (10,4,14,8), and (-1,-1,1,1). Let W be the subspace of $\mathbb{R}^4$ spanned by (1,0,0,2), (3,1,0,2), and (7,0,5,2)}. Determine the dimensions of U, W, and $U \cap W$.

7. Let U be the subspace of $\mathbb{R}^4$ spanned by {(1,0,2,0), (1,0,3,0)} and let W be the subspace of $\mathbb{R}^4$ spanned by {(1,0,0,0), (0,1,0,0), (0,0,1,1)}. Determine the dimensions of U, W, and $U \cap W$.

8. Let A be an $m \times n$ matrix with entries in F. Let U be the solution set of the homogeneous system $Ax = 0$. (This is a subspace of F^n. See Example 3 of (2.4.3).) Show that the solution set of the system $Ax = b$ is a coset of the form $v + U$ where v is one particular solution of the system $Ax = b$.

[*Hint* - see problem 6 of Problems 1B.]

2.5 **Matrices and linear maps**

Let A be an m×n matrix with entries in F where $F = \mathbb{R}$ or $F = \mathbb{C}$

We may define a linear map $g:F^n \longrightarrow F^m$ by $g(v) = Av$ for each column vector $v \in F^n$.

(The properties of matrix multiplication ensure that g is linear).

Conversely let $g:F^n \longrightarrow F^m$ be any linear map from F^n to F^m. Letting $\{e_i\},\{e'_i\}$ denote the standard bases of F^n, F^m respectively we may, for each $j = 1, 2, \ldots, n$, write $g(e_j) = \sum_{i=1}^{m} a_{ij}e'_i$ for some elements $a_{ij} \in F$, $i = 1, 2, \ldots, n$, $j = 1, 2, \ldots, m$. The m×n matrix A with entries a_{ij} is easily seen to have the property $g(v) = Av$ for each column vector $v \in F^n$.

(If v is the column vector (β_j) then we can write $v = \sum_{j=1}^{n} \beta_j e_j$ so that

$$g(v) = g(\sum_{j=1}^{n} \beta_j e_j) = \sum_{j=1}^{n} \beta_j g(e_j) = \sum_{j=1}^{n} \beta_j \sum_{i=1}^{m} a_{ij}e'_i = \sum_{i=1}^{n}\left(\sum_{j=1}^{m} a_{ij}\beta_j\right) e'_i = Av$$

by the definition of matrix multiplication.)

2.5.1 **Proposition**

There is a one-one correspondence between the set of all linear maps $F^n \longrightarrow F^m$ and the set of all m×n matrices with entries in F.

Proof

Follows easily from the above comments.

More generally let $g:V \longrightarrow W$ be a linear map between vector spaces V and W over F where dim V = n and dim W = m. Let B_1 and B_2 be bases of V and W respectively with $B_1 = \{v_1, v_2, \ldots, v_n\}$ and $B_2 = \{w_1, w_2, \ldots, w_m\}$. Since B_2 is a basis for W we may write, for each $j = 1, 2, \ldots, n$,

$g(v_j) = \sum_{i=1}^{m} a_{ij} w_i$ for some elements $a_{ij} \in F$.

The matrix $A = (a_{ij})$ is an m×n matrix called *the matrix representing g with respect to the bases B_1 and B_2*. We denote this matrix by $M_{B_1B_2}(g)$.

2.5.2 **Proposition**

Let V and W be vector spaces over F of dimension n and m respectively. There is a one-one correspondence between the set of all linear maps from V to W and the set of all m×n matrices with entries in F.

Proof

Using the above notation we associate the matrix $M_{B_1B_2}(g)$ to the linear map g. Conversely given an m×n matrix $A = (a_{ij})$ we may define a linear map $g:V \longrightarrow W$ by requiring $g(v_j) = \sum_{i=1}^{m} a_{ij} w_i$ for each basis element v_j of V and extending by linearity as in (2.3.4). This yields a one-one correspondence.

2.5.3 **Comment**

The above one-one correspondence depends on the choice of bases for V and W. A different choice of bases will yield a different one-one correspondence. Note that a choice of bases for V and W yields, via (2.3.7), isomorphisms $V \cong F^n$ and $W \cong F^m$ so that (2.5.2) can in fact be deduced from (2.5.1).

2.5.4 **Proposition**

Let B_1, B_2, and B_3 be bases for the finite-dimensional vector spaces U, V, and W . Let $f:U \longrightarrow V$ and $g:V \longrightarrow W$ be linear maps. The matrices representing f, g, and the composite g∘f with respect to the above bases are related as follows;

$$M_{B_1B_3}(g \circ f) = M_{B_2B_3}(g)\, M_{B_1B_2}(f)$$

i.e. the matrix representing the composite of two mappings is the product of the matrices representing the two individual mappings.

Proof

Let U,V, and W have dimensions n,m, and k respectively.

Let $B_1 = \{u_1, u_2, \ldots, u_n\}$, $B_2 = \{v_1, v_2, \ldots, v_m\}$, $B_3 = \{w_1, w_2, \ldots, w_k\}$.

Let $M_{B_1B_2}(f) = A = (a_{ij})$, an m×n matrix, and $M_{B_2B_3}(g) = C = (c_{ij})$, a k×m matrix.

$$\text{Then } (g\circ f)(u_j) = g(f(u_j)) = g\Big(\sum_{r=1}^{m} a_{rj} v_r\Big) = \sum_{r=1}^{m} a_{rj}\, g(v_r)$$

$$= \sum_{r=1}^{m} a_{rj} \sum_{i=1}^{k} c_{ir} w_i = \sum_{i=1}^{k} \Big(\sum_{r=1}^{m} c_{ir} a_{rj} \Big) w_i .$$

The coefficient of w_i in this last sum will be the (i,j)-entry of the matrix $M_{B_1B_3}(g\circ f)$. This coefficient is $\sum_{r=1}^{m} c_{ir} a_{rj}$ which is also the (i,j)-entry of the matrix product CA.

2.5.5 **Remarks**

The definition of matrix multiplication given in (1.1.4) may have seemed to the reader to be unnatural. The above proposition provides some motivation for that definition. When we view matrices as linear maps the product of matrices must be defined in the way given in (1.1.4) in order for composition of maps to correspond to multiplication of matrices.

We saw in (1.1.6) that matrix multiplication is associative. Proposition (2.5.4) gives an alternative proof of this since composition of functions is associative.

2.5.6 **Special cases**

Some special cases of (2.5.4) should be singled out as being of particular importance.

Firstly we may have $U = F^n$, $V = F^m$, $W = F^k$. If the bases B_1, B_2, B_3 are each the standard basis then it is easy to write down the matrices representing our linear maps. There are of course plenty of other bases for F^n, F^m, and F^k and if B_1, B_2, and B_3 are not standard bases then a little calculation is needed to determine the matrices. See problems 2D.

Secondly suppose $U = V = W$ and $B_3 = B_1$. Let 1_V denote the identity map on V. Then the matrix $M_{B_1B_2}(1_V)$ is called *the matrix of change of basis from* B_2 *to* B_1.

To determine this matrix we must write each element of B_1 as a linear combination of the elements of B_2.

Note that the matrix $M_{B_2B_1}(1_V)$, the matrix of change of basis from B_1 to B_2, is the inverse of $M_{B_1B_2}(1_V)$ because, by (2.5.4), the matrix product $M_{B_1B_2}(1_V)\, M_{B_2B_1}(1_V) = M_{B_1B_1}(1_V)$ which is clearly the identity matrix.

Thirdly let $U = V = W$, let $f = h = 1_V$, let $g:V \longrightarrow V$ be any linear map, and let $B_2 = B_1$. Then proposition (2.5.4) reads

$$M_{B_3B_3}(g) = M_{B_1B_3}(1_V)\, M_{B_1B_1}(g)\, M_{B_3B_1}(1_V)$$

If we write $A = M_{B_1B_1}(g)$, $C = M_{B_3B_3}(g)$, and $P = M_{B_3B_1}(1_V)$ then this equation is $C = P^{-1}AP$. The matrix A (respectively C) may be called the matrix of the linear map g with respect to the basis B_1 (respectively B_3).

Two matrices A and C are said to be *similar* if $C = P^{-1}AP$ for some invertible matrix P.

The above observations show that the matrices representing a linear map $g:V \longrightarrow V$ with respect to two different bases will be similar. The matrix P which gives the similarity transformation will be the matrix of change of basis.

2.5.7 **Example**

Let B be the standard basis of $\mathbb{R}^3$ and let B_1 be $\{(1,2,3),(3,2,1),(0,0,1)\}$. Show that B_1 is a basis of $\mathbb{R}^3$ and determine the two change of basis matrices $M_{BB_1}(1)$ and $M_{B_1B}(1)$. If $g:\mathbb{R}^3 \longrightarrow \mathbb{R}^3$ is given by $g(x,y,z) = (6x + y, x - y - z, 2x - y + 3z)$ determine the matrices $M_{BB}(g)$ and $M_{B_1B_1}(g)$.

It is easy to write down the matrix $M_{B_1B}(1)$ since all elements of $\mathbb{R}^3$ are naturally written in terms of the standard basis B.

Specifically $M_{B_1B}(1) = \begin{pmatrix} 1 & 3 & 0 \\ 2 & 2 & 0 \\ 3 & 1 & 1 \end{pmatrix}$.

The matrix $M_{BB_1}(1)$ is the inverse of $M_{B_1B}(1)$ and so, calculating the inverse by the methods of chapter 1, $M_{BB_1}(1) = \begin{pmatrix} -1/2 & 3/4 & 0 \\ 1/2 & -1/4 & 0 \\ 1 & -2 & 1 \end{pmatrix}$.

(We could alternatively have calculated $M_{BB_1}(1)$ directly from the definition by writing the elements of B in terms of B_1.)

The matrix $M_{BB}(g)$ can be written down immediately.

$$M_{BB}(g) = \begin{pmatrix} 6 & 1 & 0 \\ 1 & -1 & -1 \\ 2 & -1 & 3 \end{pmatrix}.$$

Now $M_{B_1B_1}(g) = M_{BB_1}(1)\, M_{BB}(g)\, M_{B_1B}(1)$ using the third special case in (2.5.6). A straightforward calculation now yields that

$$M_{B_1B_1}(g) = \begin{pmatrix} -7 & -10 & -3/4 \\ 5 & 10 & 1/4 \\ 25 & 27 & 5 \end{pmatrix}.$$

(We could alternatively have calculated $M_{B_1B_1}(g)$ directly from the definition by expressing g of each element of B_1 as a linear combination of the elements of B_1.)

2.6 **The image and kernel of a linear map**

Let $f: V \longrightarrow W$ be a linear map of vector spaces V and W.

The *image of f*, denoted Im f, is defined by $\text{Im } f = \{ f(v) : v \in V \}$

The *kernel of f*, denoted Ker f, is defined by $\text{Ker } f = \{ v \in V : f(v) = 0_W \}$

2.6.1 **Exercise**

Show that Im f is a subspace of W and that Ker f is a subspace of V.

2.6.2 **Lemma**

The linear map $f: V \longrightarrow W$ is injective if and only if Ker f = 0.

Proof

In (2.3.4) we saw that $f(0_V) = 0_W$ when f is linear. Thus if f is injective then Ker f = 0.

Conversely suppose that Ker f = 0. Then $f(v_1) = f(v_2)$ implies by linearity that $f(v_1 - v_2) = 0_W$. Hence $v_1 - v_2 = 0_V$ since Ker f = 0. This shows that f is injective.

2.6.3 **Remark**

Lemma (2.6.2) is a very useful criterion for injectivity of linear maps. To show that f is injective it suffices to show that $f(v) = 0_W$ implies $v = 0_V$.

2.6.4 **Rank and nullity of a linear map**

Let $f: V \longrightarrow W$ be a linear map of finite-dimensional vector spaces V and W. We saw in (2.6.1) that Im f and Ker f are subspaces of V and W respectively.

The dimension of Im f is called the *rank of f* and is denoted r(f).

The dimension of Ker f is called the *nullity of f* and is denoted n(f).

2.6.5 **Rank-nullity theorem**

Let $f: V \longrightarrow W$ be a linear map of finite-dimensional vector spaces V and W. Then $r(f) + n(f) = \dim V$.

Proof

Let $n(f) = n$ and let $\{u_1,u_2,\ldots,u_n\}$ be a basis of Ker f. Extend this to a basis for all of V. (This is possible by (2.2.13).)

i.e. $\{u_1,u_2,\ldots u_n,v_1,v_2,\ldots,v_s\}$ is a basis of V for suitable elements $v_1,v_2,\ldots,v_s$ of V. (If $n = 0$ we simply take a basis of V.)

We will show that $\{f(v_1),f(v_2),\ldots, f(v_s)\}$ is a basis for Im f.

First to prove that this set is linearly independent suppose that $\alpha_1 f(v_1) + \alpha_2 f(v_2) + \ldots + \alpha_s f(v_s) = 0_W$. Then the linearity of f implies that $f(\sum_{i=1}^{s} \alpha_i v_i) = 0_W$ so that $\sum_{i=1}^{s} \alpha_i v_i$ belongs to Ker f. Hence $\sum_{i=1}^{s} \alpha_i v_i$ is a linear combination of $\{u_1,u_2,\ldots u_s\}$ which implies that each $\alpha_i = 0$ because $\{u_1,u_2,\ldots,u_n,v_1,\ldots,v_s\}$ is a basis of V.

Secondly any element of Im f is a linear combination of the vectors $f(u_1),f(u_2),\ldots, f(u_n), f(v_1),\ldots,f(v_s)$. The first n of these vectors are zero as $u_1,u_2,\ldots,u_n$ are in Ker f so that $\{f(v_1),f(v_2),\ldots,f(v_s)\}$ spans Im f.

This proves that $s = r(f)$ and completes the proof.

2.6.6 **Corollary**

Let $f:V \longrightarrow W$ be a linear map and suppose $\dim V = \dim W$. Then f is injective if and only if f is surjective.

Proof

If f is injective then $n(f) = 0$ by (2.6.2) and so $r(f) = \dim V$ and f is surjective.

If f is surjective then $r(f) = \dim V$ so that $n(f) = 0$ and f is injective.

2.6.7 **Remark**

This last result is especially useful. When $\dim V = \dim W$ to prove that a linear map $f:V \longrightarrow W$ is an isomorphism it suffices to prove only one of the two properties injective and surjective, the other property following automatically. A special case of this would be when $V = W$.

2.6.8 **Example**

Show that the map $f:\mathbb{R}^4 \longrightarrow \mathbb{R}^3$, $f(w,x,y,z) = (w - x, x + y, y - z, z + x)$ is surjective.

The kernel of f is the solution set of the equations

$$w - x = 0,$$

$$x + y = 0,$$

$$y - z = 0,$$

$$z + x = 0.$$

It is easy to see that this solution set has dimension 1 so that the nullity $n(f) = 1$. The rank-nullity theorem now shows that f has rank 3 and hence f is surjective.

2.6.9 **Example**

Show that the subspace of M_nF consisting of all the $n{\times}n$ matrices with trace zero is a subspace of dimension $n^2 - 1$.

Define a map $f:M_nF \longrightarrow F$ by $f(A) =$ trace A. It is easy to see that f is surjective and so $n(f) = n^2 - 1$ by the rank-nullity theorem since F has dimension 1. This proves the result since Ker f is precisely the subspace of all matrices with trace zero.

2.6.10 **The rank of a matrix**

Let A be an $m{\times}n$ matrix with entries in F. As in (2.5) we may view A as a linear map $g:F^n \longrightarrow F^m$ where $g(v) = Av$ for each column vector $v \in F^n$. The *rank* of the matrix A is defined to be the rank of the linear map g. We write $r(A)$ for the rank of A.

Letting A_i, $i = 1,2,\ldots,n$, denote the columns of A observe that

$g(e_i) = A_i$ for each i where $\{e_i\}$ is the standard basis of F^n.

The vectors $g(e_i)$, $i = 1,2,\ldots,n$, must span Im g and hence the rank of A is the maximum number of linearly independent columns of A.

2.6.11 **Proposition**

The following are equivalent characterizations of the rank of an $m \times n$ matrix A.

(i) r(A) is the maximum number of linearly independent columns of A.

(ii) r(A) is the maximum number of linearly independent rows of A.

(iii) r(A) is the number of non-zero rows in an echelon matrix to which A has been reduced by a finite sequence of elementary row operations.

Proof

The equivalence of (i) with the definition of r(A) as the dimension of Im g follows from our comments above.

Ker g is the solution set of the homogeneous system of equations $Ax = 0$ where $x = (x_i)$ is a column vector of n unknowns. The dimension of this solution set will be $n - t$ where t is the number of non-zero rows in an echelon form for A. Now (2.6.5) tells us that $r(g) = t$. This shows that (iii) is equivalent to our definition of rank.

The nature of elementary row operations and the fact that they are all reversible implies that the subspace of F^n spanned by the rows of A coincides with the subspace of F^n spanned by the rows of an echelon form for A. Thus (ii) and (iii) are equivalent.

2.6.12 **Comment**

Another equivalent characterization of the rank of the matrix A is that $r(A) = \max\{k : \text{there exists a non-zero } k \times k \text{ minor of } A\}$. (Delete any n-k rows and any n-k columns of A. The determinant of the resulting $k \times k$ matrix is called a *k×k minor* of A.) See problem 8 of Problems 2D for a proof.

2.6.13 **Proposition**

(i) If A is any m×n matrix with entries in F where F = ℝ or ℂ then $r(A^t) = r(A)$. Also if F = ℂ then $r(\bar{A}^t) = r(A)$.

(ii) If A and B are m×n matrices with entries in F then
$r(A + B) \leq r(A) + r(B)$

(iii) If A is an m×n matrix and B is an n×p matrix, each with entries in F, then $r(AB) \leq \min(r(A), r(B))$, i.e. $r(AB) \leq r(A)$ and $r(AB) \leq r(B)$.

Proof

(i) $r(A) = r(A^t)$ by the equivalence of (i) and (ii) in (2.6.11).
It is easy to see that a set $\{v_1, v_2, .. ., v_s\}$ of vectors in $\mathbb{C}^n$ will be linearly independent if and only if $\{\bar{v}_1, \bar{v}_2, ..., \bar{v}_s\}$ is linearly independent. Hence $r(\bar{A}^t) = r(A)$.

(ii) Let $g: F^n \longrightarrow F^m$ and $h: F^n \longrightarrow F^m$ be defined by $g(v) = Av$ and $h(v) = Bv$ for each column vector $v \in F^n$. Then Im $(g + h)$ is contained in the subspace Im g + Im h as defined in (2.4.5).

The result $r(A + B) \leq r(A) + r(B)$ is a consequence of problem 4 of Problems 2C.

(iii) Let $g: F^n \longrightarrow F^m$ be given by $g(v) = Bv$ for $v \in F^n$ and let $h: F^m \longrightarrow F^p$ be given by $h(v) = Av$ for $v \in F^m$. Then $r(AB)$ equals the rank of the composite map $h \circ g$ while $r(A) = r(h)$ and $r(B) = r(g)$. Clearly Im $(h \circ g)$ is contained in Im h which shows that $r(AB) \leq r(A)$.

Clearly Ker g is contained in Ker $(h \circ g)$ and so $n(h \circ g) \geq n(g)$. Using (2.6.5) we have that $r(g) = n - n(g)$ so that $r(g) \geq n - n(h \circ g)$. Hence, by (2.6.5) again, $r(g) \geq r(h \circ g)$. This proves that $r(AB) \leq r(B)$.

Problems 2D

1. Let the vector space V be a direct sum of the subspaces $U_1, U_2, \ldots, U_n$. Let $f_i: U_i \longrightarrow U_i$ be a linear map for each $i = 1,2,\ldots,n$ and let A_i be the matrix representing f_i with respect to a basis B_i of U_i for each i. Let $f: V \longrightarrow V$ be the direct sum of the linear maps f_i.
i.e. $f(u_1 + u_2 + \ldots + u_n) = f(u_1) + f(u_2) + \ldots + f(u_n)$ for $u_i \in U_i$, $i = 1,2,\ldots,n$. Show that, with respect to the basis $B_1 \cup B_2 \cup \ldots \cup B_n$ of V f is represented by the block diagonal matrix

$$\begin{pmatrix} A_1 & & & & 0 \\ & A_2 & & & \\ & & A_3 & & \\ & & & \ddots & \\ 0 & & & & A_n \end{pmatrix}$$

2. Let $f: \mathbb{R}^3 \longrightarrow \mathbb{R}^3$ be defined by $f(x,y,z) = (x + y, 2x - y - z, x + y + z)$. Write down the matrix of f with respect to the standard basis of $\mathbb{R}^3$.
Show that $B_1 = \{(1,0,0),(1,1,0),(1,1,1)\}$ and $B_2 = \{(1,2,1),(2,1,0),(3,2,1)\}$ are each bases of $\mathbb{R}^3$.
Determine the matrices $M_{B_1B_2}(1)$, $M_{B_2B_1}(1)$, $M_{B_1B_1}(f)$, and $M_{B_2B_2}(f)$.

3. Find the rank and nullity of the linear map f in the previous problem.

4. Find the rank and nullity of the linear map $f: \mathbb{R}^4 \longrightarrow \mathbb{R}^4$ given by $f(w,x,y,z) = (w + x - 3y + z, w - x - y - z, 2w + x + 2y + z, w + 3x + 3z)$. Write down a basis for the image I of f. Determine the dimension of the subspace $I \cap U$ where U is the subspace of $\mathbb{R}^4$ spanned by $\{(1,1,2,2), (2,0,0,5)\}$.

5. Show that $B_1 = \{(1,0,0,1), (0,1,1,0), (1,0,1,0), (1,1,0,1)\}$ is a basis of $\mathbb{R}^4$. Let B denote the standard basis of $\mathbb{R}^4$ and let $g: \mathbb{R}^4 \longrightarrow \mathbb{R}^4$ be defined

by $g(w,x,y,z) = (w + x, x + y, x + 6y - 3z, w + 2x - 2z)$. Determine the matrix $M_{BB_1}(g)$.

6. Let V be the vector space of all polynomials of degree at most six. Let $g:V \longrightarrow V$ be given by $g(p(x)) = p'(x)$, the derivative of the polynomial $p(x)$. Determine the rank and nullity of the linear map g.

7. Let $f:\mathbb{R}^3 \longrightarrow \mathbb{R}^3$ and $g:\mathbb{R}^3 \longrightarrow \mathbb{R}^3$ be defined as follows ;

$$f(x,y,z) = (x + 2y + z, x - y - z, y + 3z) \text{ for all } (x,y,z) \in \mathbb{R}^3,$$

$$g(x,y,z) = (z,y,x) \text{ for all } (x,y,z) \in \mathbb{R}^3.$$

Show that $B_1 = \{(1,2,3), (1,0,1), (1,1,0)\}$ is a basis of $\mathbb{R}^3$. Determine the matrix $M_{B_1B}(g \circ f)$ where B is the standard basis of $\mathbb{R}^3$.

8. Let A be an an $m \times n$ matrix . Prove the statement in (2.6.12) that

$$\text{rank } A = \max \{ k : \text{there exists a non-zero } k \times k \text{ minor of } A \}.$$

[*Hint* - let $d = \max \{ k : \text{there exists a non-zero } k \times k \text{ minor of } A \}$.
If M is a non-zero $k \times k$ minor then the columns of M are linearly independent. Deduce that the corresponding columns of A must be linearly independent and thus $d \le r(A)$. Conversely if $r(A) = k$ then A has k linearly independent columns. Use the equivalence of (i) and (ii) in (2.6.11) to show that the $m \times k$ matrix consisting of these k linearly independent columns of A must contain a non-zero $k \times k$ minor.]

2.7 Inner products

An *inner product* on a real vector space V is a mapping $V \times V \longrightarrow \mathbb{R}$ which is symmetric, bilinear , and positive definite. We write $\langle v,w \rangle$ for the value of the inner product on the pair of vectors v, w of V.

The property of being *symmetric* means that $\langle v,w \rangle = \langle w,v \rangle$ for all $v,w \in V$.
The property of being *bilinear* means that $\langle\ \rangle$ satisfies the following two conditions ;

$<\alpha v_1 + \beta v_2, w> = \alpha <v_1,w> + \beta <v_2,w>$ for all $\alpha,\beta \in \mathbb{R}$, $v_1,v_2,w \in V$.

$<v,\alpha w_1 + \beta w_2> = \alpha <v,w_1> + \beta <v,w_2>$ for all $\alpha,\beta \in \mathbb{R}$, $v,w_1,w_2 \in V$.

The property of being *positive definite* means that $<v,v>$ is positive for all non-zero vectors $v \in V$.

An *inner product* on a complex vector space V is a mapping $V \times V \longrightarrow \mathbb{C}$ which is hermitian, sesquilinear and positive definite. We use the same notation < > as above.

The property of being *hermitian* means that $<v,w> = \overline{<w,v>}$ for all v, w in V.

The property of being *sesquilinear* means that < > satisfies the following two conditions ;

$<\alpha v_1 + \beta v_2, w> = \bar{\alpha} <v_1,w> + \bar{\beta} <v_2,w>$ for all $\alpha,\beta \in \mathbb{C}$, $v_1,v_2,w \in V$.

$<v,\alpha w_1 + \beta w_2> = \alpha <v,w_1> + \beta <v,w_2>$ for all $\alpha,\beta \in \mathbb{C}$, $v,w_1,w_2 \in V$.

The property of being *positive definite* means that $<v,v>$ is positive for all non-zero vectors $v \in V$. Note that $<v,v>$ is necessarily real-valued because of the hermitian property of < >.)

2.7.1 **Examples**

(i) Let $V = \mathbb{R}^n$ and let $x = (x_i)$, $y = (y_i)$ be vectors in $\mathbb{R}^n$.

Define $<x,y> = x_1y_1 + x_2y_2 + \ldots + x_ny_n$.

This is known as the *Euclidean inner product* or more commonly as the *dot product* on $\mathbb{R}^n$.

(ii) Let $V = \mathbb{C}^n$ and let $x = (x_i)$, $y = (y_i)$ be vectors in $\mathbb{C}^n$.

Define $<x,y> = \bar{x}_1y_1 + \bar{x}_2y_2 + \ldots + \bar{x}_ny_n$.

This is known as the *Euclidean inner product* on $\mathbb{C}^n$.

(iii) Let $V = \mathbb{R}^n$ and let $x = (x_i)$., $y = (y_i)$ be vectors in $\mathbb{R}^n$.

Let $a_1,a_2,\ldots,a_n$ be a fixed set of positive real numbers.

Define $<x,y> = a_1x_1y_1 + a_2x_2y_2 + \ldots + a_nx_ny_n$.

This is positive definite because all the a_i are positive.

2.7.2 **Definition**

The *length* or *norm* of a vector v with respect to an inner product $< \ >$ on the vector space V over F is denoted $\| v \|$ and is defined by

$$\| v \| = <v,v>^{1/2}.$$

Note that $<v,v> \in \mathbb{R}$ for all vectors $v \in V$ even when $F = \mathbb{C}$ because of the hermitian property of $< \ >$ and $<v,v> \geq 0$ for all vectors $v \in V$ because of the positive definite property of $< \ >$.

In the case of Example (i) of (2.7.1), the dot product on $\mathbb{R}^n$, the length of a vector will be our usual geometric notion of length.

If $v \neq 0_V$ then the vector λv where $\lambda = 1/\|v\|$ has unit length. The process of taking the scalar multiple λv as above is called *normalizing* v.

2.7.3 **The Cauchy-Schwarz inequality**

Let V be a vector space over F with an inner product $< \ >$.

Then $|<v,w>| \leq \|v\| \, \|w\|$ for all vectors v and w in V.

(Here $|<v,w>|$ denotes the modulus of the real or complex number $<v,w>$.)

Proof

Using the properties of inner products we see that

$$<v - \lambda w, v - \lambda w> \geq 0 \text{ for all } v,w \in V, \text{ all } \lambda \in F$$

Also $<v - \lambda w, v - \lambda w> = <v,v> - <v,\lambda w> - <\lambda w,v> + <\lambda w,\lambda w>$

$$= \|v\|^2 - \lambda <v,w> - \bar{\lambda}\overline{<v,w>} + \bar{\lambda}\lambda \|w\|^2.$$

If $w = 0_V$ then the inequality is trivially true since both sides will be zero. Thus we can assume $w \neq 0_V$.

Choosing $\lambda = <w,v>/\|w\|^2$ in our expression above we see that $\|v\|^2 - (|<v,w>|^2/\|w\|^2) \geq 0$ from which the Cauchy-Schwarz inequality follows immediately.

2.7.4 **Definition**

Let V be a vector space over F with an inner product < >.
Two vectors v and w of V are said to be *orthogonal* with respect to < > provided that $<v,w> = 0$.
The word *perpendicular* is also often used instead of orthogonal.
When $V = \mathbb{R}^n$ and < > is the dot product the above definition coincides with our usual geometric notion of orthogonality.

2.7.5 **Comment**

Our choice of λ in the proof of (2.7.3) can be given some geometric motivation as follows ;

Given v and w the vector λw where $\lambda = <w,v>/\|w\|^2$ is called the *projection of v onto w*. The vector v may be decomposed as $v = v_1 + v_2$ where $v_1 = v - \lambda w$ and $v_2 = \lambda w$, λ being as above. It is easy to check that v_1 is orthogonal to w, i.e. $<v_1,w> = 0$. Thus writing $v = v_1 + v_2$ decomposes v into two components, one in the direction of w and the other orthogonal to w. The choice of λ is the one which minimizes the length of the vector $v - \lambda w$.

2.7.6 **Definition**

A basis $\{v_1,v_2, \ldots, v_n\}$ of a finite-dimensional vector space V is called an *orthogonal basis* with respect to the inner product < > on V provided that $<v_i,v_j> = 0$ for all $i \neq j$.

The basis is called an *orthonormal basis* with respect to < > provided that it is orthogonal and in addition that $\|v_i\| = 1$ for each $i = 1,2, \ldots ,n$.

Orthonormal bases exist in any vector space V with inner product < >. The following result describes an algorithmic procedure for transforming any basis of V into an orthonormal basis.

2.7.7 The Gram-Schmidt process

Let $\{v_1,v_2,\ldots,v_n\}$ be a basis of the finite-dimensional vector space V over F. Let $<\ >$ be an inner product on V.

An orthonormal basis $\{w_1,w_2,\ldots,w_n\}$ of V is constructed as follows ;

Let $w_1 = (1/\|v_1\|)v_1$.

Let $y_2 = v_2 - (<w_1,v_2>)w_1$ and then let $w_2 = (1/\|y_2\|)y_2$.

Assuming that $w_1,w_2,\ldots,w_{k-1}$ have been defined we first let

$y_k = v_k - \sum_{i=1}^{k-1} \lambda_i w_i$ where $\lambda_i = <w_i,v_k>$ for each i and then define w_k by normalizing y_k, i.e. $w_k = (1/\|y_k\|)y_k$.

Then $\{w_1,w_2,\ldots,w_n\}$ is an orthonormal basis of V.

Proof

Note first that $\|w_i\| = 1$ for each i since the w_i are constructed by normalizing the vectors y_i. We only need to prove then that the vectors w_i are mutually orthogonal.

We prove by induction on k that the vector w_k is orthogonal to w_j for each $j = 1,2,\ldots,k-1$.

Note that if $<w_j,y_k> = 0$ then $<w_j,w_k> = 0$ so that it suffices to prove that $<w_j,y_k> = 0$ for $j = 1,2,\ldots,k-1$.

When $k = 2$, $<w_1,y_2> = <w_1,v_2 - (<w_1,v_2>)w_1>$

$$= <w_1,v_2> - <w_1,v_2><w_1,w_1> = 0.$$

i.e. the result is true for $k = 2$.

Assume the result is true for all values less than k and consider the vector w_j for j in the range $1 \le j \le k-1$. Then $<w_j,y_k> = <w_j,v_k - \sum_{i=1}^{k-1} \lambda_i w_i>$ $= <w_j,v_k> - \sum_{i=1}^{k-1} \lambda_i <w_j,w_i> = <w_j,v_k> - \lambda_j$ since by the inductive assumption

$<w_j, w_i> = 0$ for $i \neq j$. But by definition $\lambda_j = <w_j, v_k>$ so that $<w_j, y_k> = 0$ for each $j = 1,2,..,k-1$.

This shows that the result is true for k and completes the proof by induction.

2.7.8 Remark

The Gram-Schmidt process can be motivated in geometric terms via the notion of projection discussed in (2.7.5). The vector y_k is obtained by subtracting from v_k the projections of v_k onto each of the vectors $w_1, w_2, \ldots, w_{k-1}$. The resulting vector is then necessarily orthogonal to each of the vectors $w_1, w_2, \ldots, w_{k-1}$.

2.7.9 Orthogonal matrices

The $n \times n$ matrix A with real entries is said to be *orthogonal* provided that it preserves the dot product on $\mathbb{R}^n$, i.e. provided that $Ax.Ay = x.y$ for all vectors $x, y \in \mathbb{R}^n$. (The dot product was defined in (2.7.1)(i).) There are several equivalent ways of characterizing orthogonal matrices. In the following proposition the concepts such as orthogonality, length, etc. are all with respect to the dot product on $\mathbb{R}^n$.

2.7.10 Proposition

Let A be an $n \times n$ matrix with real entries.

The following statements are equivalent ;

(1) A is orthogonal.

(2) A preserves lengths, i.e. $\|Av\| = \|v\|$ for all $v \in \mathbb{R}^n$.

(3) A is invertible and $A^t = A^{-1}$, i.e. $A^tA = AA^t = I$.

(4) The rows of A form an orthonormal basis of $\mathbb{R}^n$.

(5) The columns of A form an orthonormal basis of $\mathbb{R}^n$.

Proof

(1) implies (2) because $\|v\| = (v.v)^{1/2}$ for each $v \in \mathbb{R}^n$.

Assuming (2) then $\|Av\| = \|v\|$ for all $v \in \mathbb{R}^n$.

Putting $v = x + y$ we see that $A(x + y).A(x + y) = (x + y).(x + y)$ and the properties of inner product now yield that

$$Ax.Ax + 2Ax.Ay + Ay.Ay = x.x + 2x.y + y.y$$

The definition of length of a vector, together with (2), implies that $Ax.Ay = x.y$ for all $x,y \in \mathbb{R}^n$. Thus (2) implies (1).

Next observe that $Ax.y = x.A^ty$ for all $x,y \in \mathbb{R}^n$. (To verify this check that each side equals $\sum_{i=1}^{n}\sum_{j=1}^{n} a_{ij}x_iy_j$). Using this together with (1) we have that $x.y = x.A^tAy$ for all $x,y \in \mathbb{R}^n$.

Taking $x = e_i$, $y = e_j$, the standard basis vectors, shows that the (i,j)-entry of A^tA is $e_i.e_j$ for each i,j. Hence $A^tA = I$ and A^t is the inverse of A.

Conversely if $A^tA = I$ then $x.A^tAy = x.y$ for all $x,y \in \mathbb{R}^n$ which yields that $Ax.Ay = x.y$ for all $x,y \in \mathbb{R}^n$ by using the above observation again. This shows the equivalence of (1) and (3).

The equation $AA^t = I$ amounts precisely to the fact that the rows of A form an orthonormal basis of $\mathbb{R}^n$ while the equation $A^tA = I$ amounts precisely to the fact that the columns of A form an orthonormal basis. This shows the equivalence of (3),(4), and (5).

2.7.11 **Remark**

The great value of inner products is that they enable us to exploit in general some basic geometric notions from $\mathbb{R}^2$ and $\mathbb{R}^3$, e.g. length, angle, orthogonality etc. We have seen some of this already and further applications of the inner product are given in Problems 2E below.

Problems 2E

1. Let V be the vector space of all continuous functions $f:\mathbb{R} \longrightarrow \mathbb{R}$ as defined in (2.1.2), Example 5. Show that $<f,g> = \int_{-1}^{1} f(x)g(x)dx$ is an inner product on V.

2. Let $< >$ be any inner product on $\mathbb{R}^n$. Show that $<x,y> = x^t My$ for all vectors $x,y \in \mathbb{R}^n$ where M is the symmetric $n \times n$ matrix whose (i,j)-entry is $<e_i,e_j>$, the vectors e_i being the standard basis vectors of $\mathbb{R}^n$.

Show that a similar result is valid for $\mathbb{C}^n$ with the matrix M now being hermitan.

3. Show that the length of a vector as defined in (2.7.2) has the following three properties;

(i) $\|v\| \geq 0$ and $\|v\| = 0$ if and only if $v = 0_V$.

(ii) $\|\lambda v\| = |\lambda| \|v\|$ for all $\lambda \in F$, all $v \in V$.

(iii) $\|v + w\| \leq \|v\| + \|w\|$ for all $v,w \in V$.

[*Hint* - to prove (iii) use the Cauchy-Schwarz inequality].

4. Let U be a proper subspace of the inner product space V.

Let $U^\perp = \{v \in V : <v,u> = 0 \text{ for all } u \in U\}$. Show that $U^\perp$ is a subspace of V. (It is called *the orthogonal complement of V*).

When $V = \mathbb{R}^3$ with the dot product and $U = \{\alpha(1,2,3); \alpha \in \mathbb{R}\}$ which is a line in $\mathbb{R}^3$ show that $U^\perp$ is the plane through the origin perpendicular to U.

5. Show that $V = U \oplus U^\perp$ for any subspace U of the inner product space V, $\oplus$ denoting the direct sum as defined in problem 2 of Problems 2C.

[*Hint* - choose an orthonormal basis for U, extend it to a basis for V and perform the Gram-Schmidt process on this basis.]

6. Let V be a vector space with an inner product $< >$. Let S be any non-empty subset of V and let $S^\perp = \{v \in V : <v,s> = 0 \text{ for all } s \in S\}$

Show that $S^\perp$ is a subspace of V.

Let $\mathbb{R}^n$ and $\mathbb{R}^m$ be equipped with the usual dot product and let A be an nxn matrix with real entries. Viewing A and A^t as linear maps $\mathbb{R}^n \longrightarrow \mathbb{R}^m$ and $\mathbb{R}^m \longrightarrow \mathbb{R}^n$ respectively show that Ker A = $(\text{Im } A^t)^{\perp}$ and that Im A = $(\text{Ker } A^t)^{\perp}$.

7. Let A be an nxn matrix and b a column vector in $\mathbb{R}^n$. Let x = (x_i) be a column vector of unknowns. Use the previous problem to show that only one of the following can have a solution for x;

(i) $Ax = b$

(ii) $A^t x = 0$ and $x^t b \neq 0$

(The above is sometimes known as *the Fredholm alternative.*)

8. Define the angle between two vectors v and w in an inner product space to be $\cos^{-1}(\langle v,w\rangle / \|v\| \, \|w\|)$ where $\cos^{-1}$ is the inverse cosine function.

(Thus $\langle v,w\rangle = \|v\| \, \|w\| \cos\theta$ where θ is the angle between v and w).

Show that for the dot product in $\mathbb{R}^2$ or $\mathbb{R}^3$ this coincides with the usual geometric notion of angle.

(Consider the triangle with sides v,w, and v-w and use the cosine rule).

Show that orthogonal matrices preserve angles, i.e. if A is orthogonal then the angle between Av and Aw equals the angle between v and w.

9. Show that det A = ± 1 for any orthogonal matrix A.

Show that the product of two orthogonal matrices is orthogonal.

10. Let θ be an angle, $0 \leq \theta < 2\pi$. Let $A = \begin{pmatrix} \cos\theta & -\sin\theta \\ \sin\theta & \cos\theta \end{pmatrix}$ and let

$B = \begin{pmatrix} \cos 2\theta & \sin 2\theta \\ \sin 2\theta & -\cos 2\theta \end{pmatrix}$. Show that A and B are each orthogonal matrices.

Show that A represents rotation of $\mathbb{R}^2$ in an anti-clockwise direction through an angle of θ while B represents reflection of $\mathbb{R}^2$ in the line through the origin whose slope equals $\tan\theta$.

11. Find a third column so that $\begin{pmatrix} 1/\sqrt{3} & 1/\sqrt{2} & \\ 1/\sqrt{3} & 0 & \\ 1/\sqrt{3} & -1/\sqrt{2} & \end{pmatrix}$ is orthogonal. Verify that the rows of the matrix automatically become orthonormal at the same time.

12. Use the Gram-Schmidt process to transform each of the following into an orthonormal basis ;

(i) $\{(1,1,1),(1,0,1),(0,1,2)\}$ for $\mathbb{R}^3$ with the dot product.

(ii) the same set as in (i) but using the inner product defined by $\langle (x,y,z),(x',y',z')\rangle = xx' + 2yy' + 3zz'$.

(iii) $\{1,x,x^2,x^3\}$ for the vector space of all polynomials of degree at most three using the inner product product defined by $\langle f,g\rangle = \int_{-1}^{1} f(x)g(x)dx$.

13. The $n{\times}n$ matrix A with complex entries is said to be *unitary* provided that it preserves the euclidean inner product on $\mathbb{C}^n$ as defined in (2.7.1)(ii).

Show that proposition (2.7.10) is valid for a unitary matrix A if we replace A^t by $\bar{A}^t$ everywhere in statement (3) of the proposition.

14. A linear map $f:V \longrightarrow V$ is said to be *self-adjoint* with respect to the inner product $\langle\ \rangle$ on the vector space V provided that $\langle f(x),y\rangle = \langle x,f(y)\rangle$ for all $x,y \in V$.

Let $V = \mathbb{R}^n$ and f be given by $f(x) = Ax$ for all $x \in \mathbb{R}^n$ where A is an $n{\times}n$ matrix with real entries.

Let the inner product on $\mathbb{R}^n$ be given by $\langle x,y\rangle = x^tMy$ where $M = \langle e_i,e_j\rangle$ as in problem 2 .

Show that f is self-adjoint if and only if $A^tM = MA$.

Show that $f:\mathbb{R}^3 \longrightarrow \mathbb{R}^3$, $f(x,y,z) = (y + z, -x + 2y + z, x - 3y - 2z)$ is self- adjoint with respect to the inner product $< >$ given by

$$<(x,y,z),(x',y',z')> = 2xx' + 3yy' + 2zz' + xz' + x'z + 2yz' + 2y'z$$

2.8 Miscellaneous comments

(a) The volume of a "box" in $\mathbb{R}^n$

Let $v_1,v_2,\ldots,v_n$ be a linearly independent set of vectors in $\mathbb{R}^n$. The "box" determined by $v_1,v_2,\ldots,v_n$ is defined to be the following subset of $\mathbb{R}^n$;

$$\{ \lambda_1 v_1 + \lambda_2 v_2 + \ldots + \lambda_n v_n : \lambda_i \in \mathbb{R} \text{ and } 0 \le \lambda_i \le 1 \text{ for each } i\}.$$

When $n = 2$ a "box" is a parallelogram in $\mathbb{R}^2$ and when $n = 3$ a "box" is a parallelepiped in $\mathbb{R}^3$.

We define the *volume* of the "box" determined by $v_1,v_2,\ldots,v_n$ to be $|\det(v_1\ v_2\ \ \ v_n)|$, i.e. the absolute value of the determinant of the nxn matrix with $v_1,v_2,\ldots,v_n$ as its columns.

For $n = 2$ it is an easy geometric exercise to check that the above definition gives the area of a parallelogram in $\mathbb{R}^2$.

For $n = 3$ the above definition gives the usual notion of volume of a parallelepiped in $\mathbb{R}^3$. A geometric proof of this statement requires the notion of the cross product of two vectors in $\mathbb{R}^3$. We outline this proof now, leaving the reader to check the details.

If $v = (a,b,c)$ and $w = (d,e,f)$ are vectors in $\mathbb{R}^3$ then their *cross product* is the vector $(bf - ce, cd - af, ae - bd)$ in $\mathbb{R}^3$. It is usually denoted $v \times w$.

A handy way to remember the definition of $v \times w$ is to write it as

$$\det\begin{pmatrix} e_1 & e_2 & e_3 \\ a & b & c \\ d & e & f \end{pmatrix}$$ where e_1,e_2,e_3 are the standard basis vectors of $\mathbb{R}^3$.

Orthogonality and norms used below are with respect to the usual dot product on $\mathbb{R}^3$.

The reader may check that $v \times w$ is orthogonal to both v and w and that $\|v \times w\| = \|v\| \|w\| \sin \theta$ where θ is the angle between v and w as defined in problem 8 of problems 2E.

Now consider the "box" determined by vectors u,v,and w in $\mathbb{R}^3$. Let the base of the "box" be the parallelogram determined by v and w. The reader may check that the area of this base is $\|v \times w\|$ and that the perpendicular height of the "box" is $\|u\| \cos \phi$ where ϕ is the angle between the vectors u and $v \times w$. Hence the usual notion of volume in $\mathbb{R}^3$ gives $\|v \times w\| \|u\| \cos \phi$ as the volume of the "box". But this equals the dot product $u.(v \times w)$ from problem 8 of problems 2E.

Finally the reader may check that $u.(v \times w) = \det (u\ v\ w)$, the determinant of the matrix with u,v, and w as its columns.

Let A be an n×n matrix with real entries and let $g:\mathbb{R}^n \longrightarrow \mathbb{R}^n$ be the linear map given by $g(x) = Ax$ for all $x \in \mathbb{R}^n$.

Let S be a "box" in $\mathbb{R}^n$ and write g(S) for the image under the mapping g of this "box". Then the volume of g(S) equals $|\det A|$ times the volume of S. (To prove this suppose S is determined by the vectors $v_1, v_2, \ldots, v_n$. Then g(S) is the "box" determined by $g(v_1), g(v_2), \ldots, g(v_n)$ so that the volume of g(S) equals $\det (Av_1, Av_2, \ldots, Av_n)$. The result will now follow from problem 3 of problems 1A and property (10) of (1.4.2).)

The linear map g given by the n×n matrix A thus changes volumes by a factor equal to $|\det A|$. Note that if A is orthogonal then $\det A = \pm 1$ so that an orthogonal matrix preserves volumes.

To be able to read the next two sections it will be helpful if the reader has some familiarity with the calculus of several real variables.

(b) Calculus and linear algebra

Consider first a function $f:\mathbb{R} \longrightarrow \mathbb{R}$ and suppose that the derivative of f exists at the point $x_0 \in \mathbb{R}$. Geometrically the existence of the derivative of f at x_0 amounts to the fact that a line tangential to the graph of f can be drawn at the point x_0. This line has an equation of the form $y = mx + c$ where $m = f'(x_0)$, the derivative of f at x_0, and $c = f(x_0) - mx_0$. This line can be regarded as a linear approximation to the function f at the point x_0. Strictly speaking it is not linear in the sense that we have used the word earlier unless $c = 0$ but it is the translate of a linear map by the addition of a constant c.

Now consider a map $f:\mathbb{R}^2 \longrightarrow \mathbb{R}$. The notion of f being differentiable at a point $(x_0,y_0) \in \mathbb{R}^2$ amounts to the fact that the graph of f, (which is a surface in $\mathbb{R}^3$), has a tangent plane at the point (x_0,y_0). This tangent plane has an equation of the form $ax + by - z - c = 0$ where $a = \partial f/\partial x$, $b = \partial f/\partial y$, the partial derivatives of f each evaluated at the point (x_0,y_0), and c is a constant, $c = ax_0 + by_0 - f(x_0,y_0)$. If c were zero then this tangent plane would be the graph of a linear map in the sense we have been using it while in general it will be the translate of a linear map by the addition of a constant c. As above we can regard this as a linear approximation to the function f.

Finally consider $f:\mathbb{R}^n \longrightarrow \mathbb{R}^m$ which is made up of m co-ordinate functions $f_i:\mathbb{R}^n \longrightarrow \mathbb{R}$, $i = 1,2,\ldots,m$. The definition of f being differentiable at a point $v \in \mathbb{R}^n$ amounts geometrically to the fact that there is a linear approximation to f at the point $v \in \mathbb{R}^n$, (or at least there is a translate of a linear map which approximates f at v). The $m{\times}n$ matrix of partial derivatives $\partial f_i/\partial x_j$ represents this linear map.

To summarize we may say that the differentiable functions are the ones that admit a local linear approximation.

Integal calculus of several variables also has links with some of the material we have met earlier. The definite integral of a function $f:[a,b] \longrightarrow \mathbb{R}$ over the interval $[a,b]$ on the real line can be interpreted as the area enclosed by the graph of f, the x-axis , and the lines $x = a$, $x = b$ in $\mathbb{R}^2$. The definite integral of a function $f:D \longrightarrow \mathbb{R}$, D being a subset of $\mathbb{R}^2$, can be interpreted as the volume in $\mathbb{R}^3$ lying under the graph of f and directly above the domain D of the xy-plane. Similarly the definite integral of a function of n variables can be interpreted as representing the volume of a region in $\mathbb{R}^{n+1}$.

For any $n \geq 1$ in order to evaluate a definite integral it is often necessary to make a change of variables, i.e. change from the variables $x_1,x_2,\ldots,x_n$ to a new set of variables $u_1,u_2,\ldots,u_n$. For example when $n = 2$ we sometimes change from cartesian co-ordinates x,y to polar co-ordinates r,θ and for $n = 3$ we sometimes change from cartesian co-ordinates x,y,z to spherical polar co-ordinates r,θ,ϕ.

This change of variables should be given by a differentiable function of n variables with values in $\mathbb{R}^n$, i.e. each x_i should be a differentiable real-valued function of $u_1,u_2,\ldots,u_n$. The Jacobian matrix $\partial x_i/\partial u_j$ which we met in (1.7)(b) will represent the local linear approximation to the change of variables function and its determinant will be a volume-change factor of the kind described in (2.8)(a). This indicates why it is necessary to multiply by a factor equal to $|\det(\partial x_i/\partial u_j)|$ when doing a change of variables integration of a function of several variables.

(c) Linearization

A mathematical model of a situation in science, engineering, economics, etc. may well be an equation or set of equations involving differentiable functions. One technique that is commonly used to study the model is to make a linear approximation and then apply the methods of linear algebra. The linear approximation may be obtained by taking Taylor expansions of the functions about some equilibrium point and neglecting all terms beyond the first degree terms in the expansions.

For example the linear control system described in (1.7)(e),

$$\dot{x} = Ax + Bu, \; y = Cx,$$

where $x = (x_i)$, $u = (u_i)$, $y = (y_i)$ are column vectors of length n, m, p respectively and A,B,C are matrices of size $n{\times}n$, $n{\times}m$, $p{\times}n$ respectively, may be viewed as the linearization of a non-linear control system

$$\dot{x}_i = g_i(x_1,x_2,\ldots,x_n,u_1,u_2,\ldots,u_m), \; y_i = h_i(x_1,x_2,\ldots,x_n),$$

where g_i, $i = 1, 2, \ldots, n$, and h_i, $i = 1, 2, \ldots, p$ are differentiable functions.

The matrices A,B and C arising in the linear approximation will contain the first order partial derivatives $\partial g_i/\partial x_j$, $\partial g_i/\partial u_j$, and $\partial h_i/\partial x_j$ respectively.

Chapter 3

MATRIX NORMS

The previous two chapters have been algebraic in nature. This chapter introduces some ideas from mathematical analysis that are important in matrix theory. Mathematical analysis requires a precise formulation of the notion of "nearness". In matrix analysis it will be vital to give an exact meaning to phrases such as "the matrix A is near to the matrix B" Matrix norms are the tool which will enable us to do this. They are essential in a number of different contexts. In particular they are valuable for each of the following;

(i) The definition of matrix functions such as the matrix exponential via convergent power series of matrices.

(ii) Perturbation theory for matrices, i.e. the behaviour of properties of matrices when the entries of the matrix are changed by a small amount. This topic will be dealt with in detail in chapter 7.

(iii) The analysis of the stability of algorithms used in matrix computation.

We begin by examining norms on vector spaces in general and then specialize to spaces of matrices.

3.1 Norms on vector spaces

Let V be a vector space over F where F = $\mathbb{R}$ or $\mathbb{C}$. The norm of a vector will be a measure of its size and the prototype for our general definition is the modulus of a real or complex number.

3.1.1 Definition

A *norm* on the F-vector space V is a map $V \longrightarrow \mathbb{R}$, the image of the vector v being denoted $\|v\|$, satisfying the following conditions;

(i) $\|v\| \geq 0$ for all $v \in V$, and $\|v\| = 0$ if and only if $v = 0$.

(ii) $\|\alpha v\| = |\alpha|\,\|v\|$ for all $\alpha \in F$ and all $v \in V$.

(iii) $\|v + w\| \leq \|v\| + \|w\|$ for all $v,w \in V$.

This last property (iii) is usually called "the triangle inequality".

3.1.2 Example

Let $V = \mathbb{R}^n$ and $v = (x_1,x_2,\ldots,x_n) \in \mathbb{R}^n$.

Define $\|v\| = (x_1^2 + x_2^2 + \ldots + x_n^2)^{1/2}$.

Note that $\|v\| = (v.v)^{1/2}$ where v.v is the dot product of vectors in $\mathbb{R}^n$ as defined in (2.7.1).

It is easy to see that the above definition of $\|v\|$ satisfies properties (i) and (ii) of norm. A proof of property (iii) goes as follows;

$$\|v + w\|^2 = (v + w).(v + w) = v.v + 2v.w + w.w.$$

The Cauchy-Schwarz inequality (2.7.3) says that $v.w \leq \|v\|\,\|w\|$ and hence

$$\|v + w\|^2 \leq \|v\|^2 + 2\|v\|\,\|w\| + \|w\|^2.$$

Thus $\|v + w\|^2 \leq (\|v\| + \|w\|)^2$ and taking square roots yields the desired inequality.

The above norm on $\mathbb{R}^n$ is known as the *Euclidean norm* on $\mathbb{R}^n$. The value of $\|v\|$ is the usual Euclidean distance from v to the origin in $\mathbb{R}^n$.

For $V = \mathbb{C}^n$, $v = (z_1,z_2,\ldots,z_n) \in \mathbb{C}^n$, the Euclidean norm is defined by $\|v\| = (\bar{z}_1 z_1 + \bar{z}_2 z_2 + \ldots + \bar{z}_n z_n)^{1/2}$.

3.1.3 **Example**

Let $V = F^n$ where $F = \mathbb{R}$ or $\mathbb{C}$ and $v = (x_1,x_2,\ldots,x_n) \in F^n$.

Define $\|v\| = \max(|x_1|,|x_2|,\ldots,|x_n|)$.

From the properties of modulus on F it is easy to prove that this is a norm on F^n. It is known as the *Cartesian norm* on F^n.

3.1.4 **Example**

Let $V = F^n$ where $F = \mathbb{R}$ or $\mathbb{C}$ and $v = (x_1,x_2,\ldots,x_n) \in F^n$.

Define $\|v\| = |x_1| + |x_2| + \ldots + |x_n|$.

It is easy to prove that this is a norm on F^n. It is known as the *taxicab norm* on F^n.

3.1.5 **Distance**

A norm $\| \ \|$ on a vector space V enables us to define the *distance* d(v,w) between vectors v and w by $d(v,w) = \|v - w\|$.

The Euclidean norm on $\mathbb{R}^n$ gives the usual notion of distance in $\mathbb{R}^n$.

Let V be a vector space equipped with a norm $\| \ \|$.

The *unit sphere* S in V is defined by $S = \{ v \in V : \|v\| = 1 \}$.

The shape of S will depend on the particular norm. We illustrate in Figures 3.1, 3.2, and 3.3 below the different shapes for S when $V = \mathbb{R}^2$ and $\| \ \|$ is the Euclidean, Cartesian, and taxicab norm respectively.

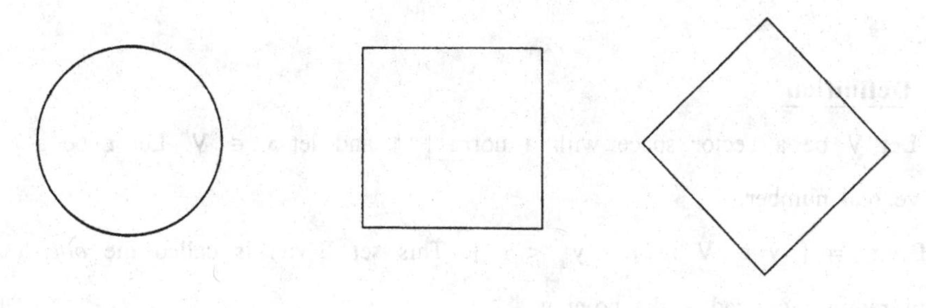

Fig. 3.1 Fig. 3.2 Fig. 3.3

3.1.6 **Norms from inner products**

Let < > be an inner product on the vector space V over F. (See (2.7) for the definition of inner product).

Define $\|v\| = (\langle v,v\rangle)^{1/2}$ for each $v \in V$.

This gives a norm on V. Properties (i) and (ii) of norm are easy to verify and property (iii) follows in the same manner as in Example (3.1.2) via the Cauchy-Schwarz inequality.

3.1.7 **Exercise**

Show that if the norm $\| \ \|$ on a vector space V comes from an inner product then $\|v + w\|^2 + \|v - w\|^2 = 2 (\|v\|^2 + \|w\|^2)$ for all v and w in V. This is known as the *parallelogram law.* When $V = \mathbb{R}^2$ with the Euclidean norm, $\|v\|$ and $\|w\|$ are the lengths of the sides of a parallelogram while $\|v + w\|$ and $\|v - w\|$ are the lengths of the diagonals of the parallelogram.

3.1.8 **Remark**

Neither the Cartesian norm nor the taxicab norm on F^n come from an inner product. See problem 2 of Problems 3A for a proof.

3.2 **Analytic concepts**

We now describe some of the important concepts from analysis that can be defined via norms.

The phrase *normed vector space* will mean a vector space equipped with a norm.

3.2.1 **Definition**

Let V be a vector space with a norm $\| \ \|$ and let $v \in V$. Let r be a positive real number.

Let $B(v;r) = \{ y \in V : \|v - y\| < r \}$. This set B(v;r) is called the *open ball* of radius r centred at the point v.

3.2.2 **Definition**

A subset S of V is an *open set* in V provided that for each element x of S there exists a positive real number r such that the open ball B(x;r) is contained in S.

(In general the real number r will depend on the point x.)

3.2.3. **Definition**

A subset T of V is a *closed set* in V provided that its complement $V \setminus T$ is an open set in V.

(The complement $V \setminus T = \{ x \in V : x \notin T \}$.)

The notions of open and closed set are thus dual notions in the following sense;

A set is open if and only if its complement is closed. A set is closed if and only if its complement is open.

See also problem 7 of Problems 3A for an equivalent definition of closed set via sequences.

Let V_1 and V_2 be vector spaces with norms $\| \ \|_1$ and $\| \ \|_2$ respectively. Let A be a non-empty subset of V_1.

3.2.4 **Definition**

A function $f : A \longrightarrow V_2$ is said to be *continuous at the point* $a \in A$ if given any positive real number ε there exists a positive real number δ such that $\| f(v) - f(a) \|_2 < \varepsilon$ whenever $\| v - a \|_1 < \delta$.

We say that f is *continuous* if f is continuous at each point $a \in A$.

(This definition mimics the usual definition of continuity of a real-valued function of one real variable, with the norm playing the role of the modulus of a real number.)

3.2.5 **Definition**

A sequence $\{v^{(n)}\}$ of vectors in a normed vector space V is said to *converge* to the element $v \in V$ if, given any positive real number ε, there exists a natural number n_0 such that $\|v^{(n)} - v\| < \varepsilon$ for all $n > n_0$.

3.2.6 **Definition**

A sequence $\{v^{(n)}\}$ of vectors in a normed vector space V is said to be a *Cauchy sequence* if, given any positive real number ε, there exists a natural number n_0 such that $\|v^{(n)} - v^{(m)}\| < \varepsilon$ for all $n > n_0$, $m > n_0$.

(Definitions (3.2.5) and (3.2.6) mimic the usual definitions for sequences of real numbers.)

3.2.7 **Exercise**

Show that if a sequence of vectors converges then it must be a Cauchy sequence.

3.2.8 **Definition**

A normed vector space V is *complete* if every Cauchy sequence of vectors in V converges to an element of V.

3.2.9 **Theorem**

Let $V = \mathbb{R}$ or $\mathbb{C}$ with the usual modulus of a real or complex number as the norm. Then V is complete.

Proof

This is a standard result in mathematical analysis and is usually known as the *Cauchy convergence criterion.* A proof can be found in most textbooks on real or complex analysis. See [R] for example.

3.2.10 **Definition**

Two norms $\| \ \|_1$ and $\| \ \|_2$ on a vector space V over $\mathbb{R}$ or $\mathbb{C}$ are *equivalent* if there exist positive real numbers k_1 and k_2 such that $\|v\|_1 \leq k_1 \|v\|_2$ and $\|v\|_2 \leq k_2 \|v\|_1$ for all $v \in V$.

The motivation for the above definition is that equivalent norms will be the same from the point of view of continuity and convergence. For example, if $\| \ \|_1$ and $\| \ \|_2$ are norms on a vector space V then the sequence $v^{(n)}$ of vectors in V will converge to the vector $v \in V$ with respect to $\| \ \|_1$ if and only if it converges to v with respect to $\| \ \|_2$.

3.2.11 **Theorem**

Let V be a finite-dimensional vector space over $\mathbb{R}$ or $\mathbb{C}$. All norms on V are equivalent.

Proof

See the appendix to this chapter.

The significance of this theorem is that for problems involving convergence and continuity it does not matter which norm on V is used.

3.2.12 **Theorem**

Any finite-dimensional normed vector space is complete.

Proof

If V is a k-dimensional vector space over F, F = $\mathbb{R}$ or $\mathbb{C}$, then V is isomorphic to F^k by (2.3.7). In view of (3.2.11) it suffices to prove the theorem in the case when $V = F^k$ equipped with the Cartesian norm.

Let $\{v^{(n)}\}$ be a Cauchy sequence in F^k and write $v^{(n)}$ as a k-tuple $(v_1^{(n)}, v_2^{(n)}, \ldots, v_k^{(n)})$ of elements of F.

Given $\varepsilon > 0$ there exists a natural number n_0 such that for all $n,m > n_0$ we have $\|v^{(n)} - v^{(m)}\| < \varepsilon$ where $\| \ \|$ is the Cartesian norm. This means that $\|v^{(n)}\| = \max |v_i^{(n)}|$ for each vector $v^{(n)}$. The sequence of i-th co-ordinates $\{v_i^{(n)}\}$ will thus be a Cauchy sequence in F for each i = 1,2,....,n. F is complete by (3.2.9) so that the sequence $\{v_i^{(n)}\}$ will converge to an element of F which we will denote by v_i. Letting $v = (v_1, v_2, \ldots, v_k) \in F^k$ we see that $\|v^{(n)} - v\| = \max \|v_i^{(n)} - v_i\|$ which will tend to zero as n tends to infinity because each $v_i^{(n)}$ converges to v_i. This shows that the sequence $\{v^{(n)}\}$ converges to v in F^k and completes the proof.

3.2.13 **Lemma**

Let $\{v^{(n)}\}$ be a sequence of vectors in a finite-dimensional normed vector space V. If the sequence $\{\|v^{(n)}\|\}$ converges in $\mathbb{R}$ then the sequence $\{v^{(n)}\}$ converges in V.

Proof

As in (3.2.12) we may assume $V = F^k$ with the Cartesian norm. We will use the same notation as in (3.2.12).

If the sequence $\{\|v^{(n)}\|\}$ converges in $\mathbb{R}$ then it is a Cauchy sequence in $\mathbb{R}$ by (3.2.7). Hence, for each $i = 1,2,\ldots,k$, the sequence $\{v_i^{(n)}\}$ is a Cauchy sequence in F and, by (3.2.9), this converges to some element v_i of F. Letting $v = (v_1,v_2,\ldots,v_k)$ it is easy to see that $\{v^{(n)}\}$ converges to $v \in F^k$ with respect to the Cartesian norm.

3.2.14 **Series of vectors**

Let $\{v_n\}$ be a sequence of vectors in a normed vector space V. We may form the *n-th partial sum* $s_n = v_1 + v_2 + \ldots + v_n$.

The infinite series $\sum_{i=1}^{\infty} v_i = v_1 + v_2 + \ldots + v_n + \ldots\ldots$

is said to *converge* to the vector $v \in V$ provided that the sequence of partial sums s_n converges to v.

3.2.15 **Exercise**

Let $\{v_n\}$ be a sequence of vectors in a finite-dimensional F-vector space V. Show that if $\sum_{i=1}^{\infty} \|v_i\|$ is a convergent series in $\mathbb{R}$ then $\sum_{i=1}^{\infty} v_i$ converges to some vector $v \in V$. [*Hint* - use (3.2.13).]

3.2.16 **Remark**

Beware that (3.2.15) need not be true if V fails to be finite-dimensional. It is only true if V is complete.

Problems 3A

1. Let V be any normed vector space. Show that the norm mapping $f:V \longrightarrow \mathbb{R}$, $f(v) = \|v\|$ for all $v \in V$, is continuous.

[*Hint* - write $u = (u - v) + v$ and use the triangle inequality to prove that $|\|u\| - \|v\|| \leq \|u - v\|$ for all $u,v \in V$.]

2. Use the parallelogram law (3.1.7) to prove that neither the Cartesian norm nor the taxicab norm come from an inner product.

3. Let $\| \|_e$, $\| \|_c$, and $\| \|_t$ denote the Euclidean, Cartesian, and taxicab norms respectively on F^n, $F = \mathbb{R}$ or $\mathbb{C}$. Show that each of the following identities is valid for all $v \in F^n$.

$$\|v\|_c \leq \|v\|_t$$

$$\|v\|_t \leq n\ \|v\|_c$$

$$\|v\|_e \leq \|v\|_t$$

$$\|v\|_t \leq (n)^{1/2}\ \|v\|_e$$

$$\|v\|_c \leq \|v\|_e$$

$$\|v\|_e \leq (n)^{1/2}\ \|v\|_c$$

4. Let $\| \|$ be any norm on F^n. The *dual norm* $\| \|_D$ of $\| \|$ is defined by $\|v\|_D = \max(|\bar{v}^t z|)$, maximum over all $z \in F^n$ with $\|z\| = 1$.

Show that $\| \|_D$ is a norm on F^n. Show that the Euclidean norm is its own dual. Show that the Cartesian and taxicab norms are duals of each other.

5. Let the function $f:A \longrightarrow V_2$ be as in definition (3.2.4). Show that f is continuous at the point $a \in A$ if and only if for each sequence $\{a_n\}$ in A which converges to the point a the sequence of values $f(a_n)$ converges to

the value f(a).

6. Let f be as in problem 5. Show that f is continuous if and only if $f^{-1}(U)$ is an open set in A for each open set U in V_2.

($f^{-1}(U) = \{ x \in A : f(x) \in U \}$.)

(This result is known as the *open-set-characterization of continuity.*)

7. Let S be a subset of the normed vector space V. Prove that S is a closed set if and only if each sequence of points of S which converges in V in fact converges to a point of S.

8. Let $< >$ be an inner product on a vector space V. Let $\{x_n\}$ and $\{y_n\}$ be sequences in V converging to the points x and y respectively. Prove that $<x_n,y_n>$ converges in to $<x,y>$.

(This result can be interpreted as saying that any inner product is continuous!)

(*Hint* - write $<x_n,y_n> - <x,y> = <x_n,y_n> - <x_n,y> + <x_n,y> - <x,y>$ and use the triangle inequality and the Cauchy-Schwarz inequality.)

3.3 Norms on spaces of matrices

The set of all m×n matrices over F forms an F-vector space of dimension mn in a natural way. Many different norms can be defined on this space and, by (3.2.11), they are all equivalent. We may thus use whichever norm happens to be most convenient for the problem in hand.

3.3.1 **Example** *The Euclidean norm*

$$\|A\| = (\sum_{i=1}^{m} \sum_{j=1}^{n} |a_{ij}|^2)^{1/2} \text{ where } A = (a_{ij}) \text{ is an m×n matrix.}$$

This is also sometimes called the *Frobenius norm on matrices.*

3.3.2 **Example** *The Cartesian norm*

$$\|A\| = \max_{i,j} |a_{ij}| \text{ where } A = (a_{ij}) \text{ is an m×n matrix.}$$

3.3.3 **Example** *The taxicab norm*

$$\|A\| = \sum_{i=1}^{m}\sum_{j=1}^{n} |a_{ij}| \text{ where } A = (a_{ij}) \text{ is an } m\text{x}n \text{ matrix.}$$

3.3.4 **Example** The *maximum absolute row sum norm*

$$\|A\| = \max_{i}\left(\sum_{j=1}^{n} |a_{ij}|\right) \text{ where } A = (a_{ij}) \text{ is an } n\text{x}n \text{ matrix.}$$

3.3.5 **Example** The *maximum absolute column sum norm*

$$\|A\| = \max_{j}\left(\sum_{i=1}^{n} |a_{ij}|\right) \text{ where } A = (a_{ij}) \text{ is an } n\text{x}n \text{ matrix.}$$

3.3.6 **Definition**

A norm on the space of all nxn matrices M_nF, $F = \mathbb{R}$ or $\mathbb{C}$, is said to satisfy *property (*)* if $\|AB\| \le \|A\| \, \|B\|$ for all A and B in M_nF.

Matrix norms which have this extra property (*) are especially useful. An important class of such norms are the *operator norms* which we will define now.

Let $A \in M_nF$ and view A as the linear map $F^n \longrightarrow F^n, v \longrightarrow Av$. Let $\| \ \|$ be any norm on F^n.

3.3.7 **Lemma**

$\{ \|Av\|/\|v\| : v \in F^n, v \neq 0 \}$ is a bounded subset of $\mathbb{R}$.

(i.e. the set of ratios of the norm of Av to the norm of v is a bounded set.)

Proof

Let $\{e_i\}$, $i = 1,2, \ldots, n$, be the standard basis of F^n. Each element $v \in F^n$ can be written as $v = \sum_{i=1}^{n} v_i e_i$ for suitable scalars $v_i \in F$.

Then $\|Av\| = \|\sum_{i=1}^{n} v_i Ae_i\|$ using the linearity of A,

$\le \sum_{i=1}^{n} |v_i| \, \|Ae_i\|$ using properties of norm,

$\le (\max_i |v_i|) \left(\sum_{i=1}^{n} \|Ae_i\| \right)$.

Now $\max|v_i|$ is the Cartesian norm of v and, by (3.2.11), it is equivalent to $\| \ \|$. Hence there exists a positive real number k such that $\max |v_i| \le k\|v\|$.

This shows that $\|Av\|/\|v\| \le k(\sum_{i=1}^{n}\|Ae_i\|)$ and this completes the proof because the right-hand side of this inequality is a finite constant which is independent of the vector v.

3.3.8 **Definition**

The *operator norm* of the n×n matrix A is the supremum, i.e. least upper bound, of $\{ \|Av\|/\|v\| : v \in F^n, v \ne 0 \}$.

$$\|A\| = \sup \{ \|Av\|/\|v\| : v \in F^n, v \ne 0 \}$$

This set is bounded by (3.3.7) and it is a basic axiom of the real number system that any bounded subset of $\mathbb{R}$ has a least upper bound. Note that, without (3.3.7), our definition would not be meaningful.

3.3.9 **Remark**

Equivalently we can define the operator norm as follows;

$$\|A\| = \sup \{ \|Av\| : v \in F^n \text{ and } \|v\| = 1 \}$$

3.3.10 **Exercise**

Show that the definitions in (3.3.8) and (3.3.9) are equivalent.

(Note that $v/\|v\|$ has norm one !)

3.3.11 **Lemma**

The operator norm is a norm which satisfies property (*).

Proof

Properties (i) and (ii) in the definition of norm are immediate from (3.3.8). Property (iii) of norm follows from the equivalent definition (3.3.9) and the triangle inequality for the underlying norm on F^n.

To prove property (*) let A and B be n×n matrices. Using (3.3.8)

$$\|AB\| = \sup \{ \|ABv\|/\|v\| : v \in F^n \text{ and } \|v\| \neq 0 \}$$
$$= \sup \{ (\|ABv\|/\|Bv\|)(\|Bv\|/\|v\| : v \in F^n \text{ and } Bv \neq 0 \}.$$
$$\leq \|A\| \, \|B\|.$$

(Note that $Bv = 0$ implies that $ABv = 0$ so that restricting to the set of v for which $Bv \neq 0$ will not change the supremum. We have also used the fact that the supremum of a product is less than or equal to the product of the suprema.)

3.3.12 **Comments on the operator norm**

(i) The operator norm is defined with respect to a particular choice of norm on F^n. For a different choice of norm on F^n we will obtain a different operator norm on M_nF.

(ii) An operator norm of $A \in M_nF$ measures by how much the linear map on F^n given by A magnifies the unit ball B(0:1).

i.e. $\|A\|$ is the radius of the smallest ball in F^n within which we can fit the image under A of the ball B(0:1).

(iii) The name "operator norm " arises from the fact that definition (3.3.8) makes sense in the following more general context:

Let V be any normed vector space and let $g : V \longrightarrow V$ be any linear operator on V. If $\{ \|g(v)\|/\|v\| : v \in V, v \neq 0\}$ is a bounded subset of $\mathbb{R}$ we say that g is a *bounded* operator.

Definition (3.3.8) can be used to define $\|g\|$. Note that (3.3.7) says that any linear operator on a finite-dimensional vector space is bounded. On infinite-dimensional vector spaces there can exist unbounded linear operators.

(iv) From the definition of operator norm we have the property that $\|Av\| \leq \|A\| \, \|v\|$ for all vectors $v \in F^n$.

3.4 Examples and properties of operator norms

If we start with a familiar norm such as the Euclidean, Cartesian, or taxicab norm on F^n we may ask what is the corresponding operator norm on M_nF.

3.4.1 Proposition

(a) If $\| \ \|$ is the Cartesian norm on F^n then the corresponding operator norm on M_nF is the maximum absolute row sum norm as defined in (3.3.4).

(b) If $\| \ \|$ is the taxicab norm on F^n then the corresponding operator norm on M_nF is the maximum absolute column sum norm as defined in (3.3.5).

Proof

(a) Let $\| \ \|_c$ be the Cartesian norm on F^n and let $\| \ \|$ be the resulting operator norm on M_nF. Let $A \in M_nF$, $A = (a_{ij})$, and let v be a column vector in F^n. Write $v = (v_i)$ for $v_i \in F$, $i = 1,2, ..,n$.

$$Av = \begin{pmatrix} a_{11} & a_{12} & \cdots & a_{1n} \\ a_{21} & a_{22} & \cdots & a_{2n} \\ & & & \\ & & & \\ a_{n1} & a_{n2} & \cdots & a_{nn} \end{pmatrix} \begin{pmatrix} v_1 \\ v_2 \\ \\ \\ v_n \end{pmatrix} = \begin{pmatrix} \sum_j a_{1j}v_j \\ \sum_j a_{2j}v_j \\ \\ \\ \sum_j a_{nj}v_j \end{pmatrix}$$

From the definition of $\| \ \|_c$ and the properties of modulus we see that

$$\|Av\|_c = \max_i(|\sum_j a_{ij}v_j|) \le \max_i (\sum_j |a_{ij}||v_j|) \le \|v\|_c \max_i (\Sigma|a_{ij}|)$$

Thus $\|Av\|_c \le \|v\|_c\|A\|_2$ where $\| \ \|_2$ denotes the maximum absolute row sum norm. The definition of operator norm now implies that $\|A\| \le \|A\|_2$.

To prove equality of $\|A\|$ and $\|A\|_2$ we will produce a vector v in F^n for which $\|Av\|_c / \|v\|_c = \|A\|_2$. Let the maximum absolute row sum for the matrix A occur for the i-th row of A. The choice of i need not be unique. We select one such i and keep it fixed.

Define the vector $v = (v_j)$ by $v_j = \frac{|a_{ij}|}{a_{ij}}$ if $a_{ij} \neq 0$,

$$v_j = 0 \text{ if } a_{ij} = 0.$$

It is easy to see that $\|v\|_c = 1$ and $\|Av\|_c = \|A\|_2$. This completes the proof of (a).

The proof of (b) goes similarly and we leave the details as an exercise for the reader.

3.4.2 **Remark**

The above proposition shows in particular that both the maximum absolute row and column sum norms are operator norms and each must thus satisfy property (*) of (3.3.6).

A description of the operator norm arising from the Euclidean norm on F^n is deferred until (6.8.7).

3.4.3 **Lemma**

Let I denote the identity matrix in M_nF. Then $\|I\| = 1$ for any operator norm $\| \; \|$ on M_nF.

Proof

$\|Iv\|/\|v\| = 1$ for all $v \neq 0$ since $Iv = v$. Thus $\|I\| = 1$.

Corollary

The Euclidean norm on M_nF is not an operator norm for any $n > 1$.

Proof

$\|I\| = (n)^{1/2}$ for the Euclidean norm.

It is also true that neither the Cartesian nor the taxicab norms on M_nF are operator norms for any $n > 1$. See problem 1 of Problems 3B.

3.4.5 **Remark**

Some norms which are not operator norms may nevertheless satisfy property (*). For example the Euclidean norm on M_nF satisfies property (*). See problem 2 of Problems 3B.

3.4.6 **Exercise**

Show that $\|A^n\| \leq \|A\|^n$ for each natural number n provided that $\| \ \|$ satisfies property (*). This inequality is especially useful in the study of power series of matrices.

Problems 3B

1. Prove that the Cartesian and taxicab norms are not operator norms on M_nF.
2. Prove that the Euclidean norm on M_nF satisfies property (*) of (3.3.6).

3.5 Matrix functions defined by power series

Given some function f(x) of a real or complex variable one may ask whether replacing x by the matrix A is possible, i.e. can any meaning be given to f(A) ?

When A is a square matrix and f is a power series, i.e. $f(x) = \sum_{k=0}^{\infty} c_k x^k$ where each $c_k \in \mathbb{R}$ or $\mathbb{C}$, it is possible to give meaning to f(A) provided that certain convergence conditions are satisfied. (Note-we define $A^0 = I$.)

3.5.1 **Proposition**

Let $A \in M_n\mathbb{C}$ and $f(x) = \sum_{k=0}^{\infty} c_k x^k$, a power series in one variable x, the coefficients c_k being in $\mathbb{R}$ or $\mathbb{C}$. Then the power series of matrices $\sum_{k=0}^{\infty} c_k A^k$ will converge to some matrix in $M_n\mathbb{C}$ provided that the series $\sum_{k=0}^{\infty} c_k \|A\|^k$ converges in $\mathbb{R}$ for some norm $\| \ \|$ on $M_n\mathbb{C}$ which satisfies $\|A^k\| \leq \|A\|^k$ for all natural numbers k.

(Note - this last condition on the norm will hold for all norms satisfying property (*) of (3.3.6) and in particular for all operator norms.)

Proof

Let $s_m = \sum_{k=0}^{m} c_k A^k$. It suffices, by (3.2.13), to show that $\{s_m\}$ is a convergent sequence in $\mathbb{R}$. Now $\|s_n - s_m\| = \|\sum_{k=m+1}^{n} c_k A^k\|$ for any $n > m$ and then, via properties of norm including the extra one stated in the proposition, we see that $\|s_n - s_m\| \leq \sum_{k=m+1}^{n} |c_k| \, \|A\|^k$. Since $\sum_{k=0}^{\infty} c_k \|A\|^k$ converges in $\mathbb{R}$ by the assumption in the proposition, it follows from (3.2.7) that it is a Cauchy sequence in $\mathbb{R}$. Hence $\{s_m\}$ is a Cauchy sequence in $\mathbb{R}$ and, by (3.2.9), it converges in $\mathbb{R}$.

3.5.2 **Example**

Recall that the *exponential function* $e^x = \sum_{k=0}^{\infty} x^k/k!$. We may thus define the *matrix exponential,* denoted exp A or e^A for any square matrix A, to be the sum of the convergent power series of matrices $\sum_{k=0}^{\infty} A^k/k!$. (Note that (3.5.1) ensures the convergence of this power series because $\sum_{k=0}^{\infty} \|A\|^k/k!$ converges in $\mathbb{R}$ to $e^{\|A\|}$.)

3.5.3 **Comment**

The matrix exponential occurs in many applications, especially in control theory problems in engineering. We will see in chapter 5 that the matrix exponential appears naturally in solving a system of linear differential equations.

3.5.4 **Comment**

The calculation of the matrix exponential is especially simple when the matrix A is similar to a diagonal matrix. If $S^{-1}AS = D$ for some invertible matrix S and D is diagonal with entries $\lambda_1, \lambda_2, \ldots, \lambda_n$ then $e^A = S(e^D)S^{-1}$ where e^D is diagonal with entries $e^{\lambda_1}, e^{\lambda_2}, \ldots, e^{\lambda_n}$. (This is because

$A = SDS^{-1}$ implies that $A^k = SD^kS^{-1}$ for each k.)

We will see more about the matrix exponential in chapter 5 .

3.5.5 **Comment**

Using their convergent power series expansions we may define trigonometric functions such as sin A and cos A for a square matrix A, also hyperbolic functions such as sinh A and cosh A.

Problems 3C

1. Let $A \in M_nF$, $B \in M_nF$, and suppose that $AB = BA$.
Show that $\exp(A+B) = (\exp A)(\exp B)$.

2. Show that the conclusion of problem 1 is false in general if A and B do not commute.
(*Hint* - let $A = \begin{pmatrix} 0 & 1 \\ 0 & 0 \end{pmatrix}$, $B = \begin{pmatrix} 0 & 0 \\ 1 & 0 \end{pmatrix}$, and show that $\exp(A+B)$ and $(\exp A)(\exp B)$ have different traces.)

3. If $A = \begin{pmatrix} 1 & 0 \\ 1 & 0 \end{pmatrix}$ show that $\sin A = \begin{pmatrix} \sin 1 & 0 \\ \sin 1 & 0 \end{pmatrix}$.

4.(i) Show that $\exp 0 = I$ where 0 denotes the n×n matrix with all entries zero and I denotes the identity n×n matrix.

(ii) Show that exp A is invertible for any n×n matrix A and that exp (-A) is its inverse.

3.6 **A criterion for invertibility**

We now give a criterion in terms of norms for a matrix to be invertible. It is often known as the *Banach lemma*, named after the Polish mathematician Banach.

3.6.1 **Banach lemma**

Let $B \in M_nF$ and let $\| \ \|$ be any operator norm on M_nF.
If $\|B\| < 1$ then $I + B$ is invertible, and the following inequality holds ;

$$(1 + \|B\|)^{-1} \le \| (I + B)^{-1} \| \le (1 - \|B\|)^{-1}.$$

Proof

Assume that $I + B$ is not invertible. Then the linear map given by $I + B$ has non-zero kernel so that $(I + B)v = 0$ for some non-zero vector $v \in F^n$. Thus $Bv = -v$ which implies that $\|B\| \geq 1$ by the definition of operator norm. This contradicts our initial assumption that $\|B\| < 1$ and so $I + B$ must be invertible.

Using properties of operator norms including property (*) we see that

$$1 \leq \|I\| \leq \|(I + B)(I + B))^{-1}\| \leq \|(I + B)\| \ \|(I + B)^{-1}\|$$

$$\leq (\|I\| + \|B\|)(\|(I + B)^{-1}\|).$$

Since $\|I\| = 1$ this yields $(1 + \|B\|)^{-1} \leq \| (I + B)^{-1}\|$ and the first half of the inequality is proved.

For the second half we write $I = (I + B)(I + B)^{-1}$

$$= I(I + B)^{-1} + B(I + B)^{-1}.$$

Now $(I + B)^{-1} = I - B(I + B)^{-1}$ and, via properties of norms, we have

$$\|(I + B)^{-1}\| = \|I - B(I + B)^{-1}\| \leq \|I\| + \| B(I + B)^{-1}\|$$

$$\leq 1 + \|B\| \ \|(I + B)^{-1}\|.$$

Re-arranging gives $(1 - \|B\|)(\|(I + B)^{-1}\| \leq 1$ and this completes the proof.

3.6.2 **Comment**

Consider the following geometric series;

$$(1 + x)^{-1} = 1 - x + x^2 - x^3 . . . . + (-1)^n x^n + ...$$

Using this together with (3.5.1) we obtain an infinite series expansion $(I + B)^{-1} = \sum_{n=0}^{\infty} (-1)^n B^n$ which is a convergent power series of matrices provided that $\|B\| < 1$.

Similarly $(I - B)^{-1} = \sum_{n=0}^{\infty} B^n$ for $\|B\| < 1$ via the series for $(1-x)^{-1}$.

Problems 3D

1. Let $A = (a_{ij})$ be the 100x100 matrix with $a_{ij} = 1$ for all $i = j$, and with $a_{ij} = 10^{-2}$ for all $i \neq j$. Use the Banach lemma to show that A is invertible.
2. Let A and B be nxn matrices with A invertible but B not invertible. Show that $\|A - B\| \geq (\|A^{-1}\|)^{-1}$ for any operator norm.
(*Hint* - write $B = A(I - Y)$ for a suitable matrix Y.)

3.7 **Stability of algorithms**

In practical matrix problems scientists and engineers often find it necessary to solve the problem with a computer using an appropriate algorithm. (Solving a system of linear equations involving a large number of variables is one such example.) The computed solution will be inaccurate to some extent due to rounding errors on the computer. Also there could be some inaccuracy in the data if it has come from experimental measurements. To compare the computed solution with the true solution some kind of quantitative error analysis is needed. Matrix norms are an essential tool in the analysis of algorithms used in matrix computation.

An algorithm is said to be *stable* provided that the computed solution via the algorithm is the exact solution of a slightly perturbed version of the original problem. For example, suppose that the problem is to solve the system of linear equations $Ax = b$ where x is the column vector of unknowns, $A \in M_n\mathbb{R}$, and $b \in \mathbb{R}^n$. An algorithm producing a solution $\hat{x}$ is stable provided that $A_1\hat{x} = b_1$ for some $A_1 \in M_n\mathbb{R}$, $b_1 \in \mathbb{R}^n$ with $\|A - A_1\|$ and $\|b - b_1\|$ each being very small. The method of Gaussian elimination for systems of linear equations, as described in chapter 1, can be shown not to yield a stable algorithm. However a slight refinement, known as *Gaussian elimination with complete pivoting*, does turn out to be stable.

Whenever a small change in the data of a numerical problem leads to a relatively larger change in the numerical values of the solution the problem is said to be *ill-conditioned.* If the change in solution is not of greater order than the change in the data then the problem is said to be *well-conditioned.* A stable algorithm applied to a well-conditioned problem should yield a solution very close to the exact solution. A stable algorithm applied to an ill-conditioned problem should introduce no further errors beyond those already present due to the inaccuracy of the data.

A detailed treatment of all this goes well beyond the scope of this book. The book [GVL] is recommended to the reader who is interested in numerical matrix computation.

APPENDIX TO CHAPTER 3

A proof of the equivalence of all norms on $\mathbb{R}^n$ and $\mathbb{C}^n$

A subset S of $\mathbb{R}^n$ or $\mathbb{C}^n$ is said to be *compact* if and only if each sequence of points in S has a subsequence which converges to a point of S. At the moment all references to convergence are with respect to the Euclidean norm on $\mathbb{R}^n$ or $\mathbb{C}^n$, i.e. the usual notion of distance.

A subset S of $\mathbb{R}^n$ or $\mathbb{C}^n$ is said to be *bounded* if S is contained in an open ball B(0;r) for some positive real number r.
It can be shown that a subset of $\mathbb{R}^n$ or $\mathbb{C}^n$ is compact if and only if it is both closed and bounded.

(We should remark that there is a more general notion of compactness defined via open sets but for our purposes the above definition will suffice.)

Lemma 1

Let $f : S \longrightarrow \mathbb{R}$ be a continuous function defined on a compact subset S of F^n, $F = \mathbb{R}$ or $\mathbb{C}$. Then f attains both a maximum and a minimum value on S.

Proof

We first show that f(S) is compact. Let $\{y_n\}$ be a sequence of points in f(S) so that $y_n = f(x_n)$ for points x_n in S. Since S is compact there exists a subsequence $\{x_{n_k}\}$ of $\{x_n\}$ which converges to a point $x \in S$. Since f is continuous $f(x_{n_k})$ converges to f(x). This shows that f(S) is compact.

Then f(S) must be a closed and bounded subset of $\mathbb{R}$. It is a fundamental axiom of the real number system that any such set has a supremum, i.e. least upper bound. Also the set S must have an infimum, i.e. greatest lower bound. The supremum and infimum of S must necessarily lie in S because S is a closed set. This proves the lemma.

Lemma 2

Let $\| \ \|$ be any norm on F^n. Then the mapping $g : F^n \longrightarrow \mathbb{R}$, given by $g(x) = \|x\|$ for each $x \in F^n$ is a continuous function with respect to the Euclidean norm on F^n.

Proof

By applying the triangle inequality to (x - y) + y we see that

$|(\|x\| - \|y\|)| \leq \|x - y\|$ for all vectors $x,y \in F^n$.

Let $\| \ \|_0$ denote the Euclidean norm on F^n and let $\{e_1,e_2,..,e_n\}$ denote the standard basis of F^n.

Note that $|x_i| \leq \|x\|_0$ for all i as $\|x\|_0 = (|x_1|^2 + .. + |x_n|^2)^{1/2}$.

$$\text{Hence } \|x\| = \| \sum_{i=1}^{n} x_i e_i \| \leq \sum_{i=1}^{n} |x_i| \|e_i\| \leq \|x\|_0 \left(\sum_{i=1}^{n} \|e_i\| \right).$$

Thus $\|x\| \leq k \|x\|_0$ for all $x \in F^n$ where $k = \sum_{i=1}^{n} \|e_i\|$, and from this it follows that $|(\|x\| - \|y\|)| \leq \|x - y\| \leq k\,\|x - y\|_0$ for all x and y in F^n. Now, given any $\varepsilon > 0$, we take $\delta = \varepsilon/k$ and we have that $|(\|x\| - \|y\|)| < \varepsilon$ whenever $\|x - y\|_0 \leq \delta$. This proves the lemma.

Theorem

All norms on F^n are equivalent where $F = \mathbb{R}$ or $\mathbb{C}$.

Proof

We must show that any two norms $\|\ \|_1$ and $\|\ \|_2$ on F^n satisfy the inequalities in (3.2.10).

Let $S = \{ x \in F^n : \|x\|_0 = 1\}$ where $\|\ \|_0$ is the Euclidean norm. S is a compact set as it is closed and bounded. Define a function $f : S \longrightarrow \mathbb{R}$ by $f(x) = (\|x\|_1)/(\|x\|_2)$ for each $x \in S$. Note that $\|x\|_2 \neq 0$ for any $x \in S$ and note also that f is positive-valued. By lemma 2, f is continuous and then, by lemma 1, there exist positive constants c and d such that $c \leq f(x) \leq d$ for all $x \in S$.

Any $v \in F^n$ is expressible as $v = \alpha x$ for some $x \in S$, $\alpha \in F$. (Specifically $\alpha = \|v\|_0$ and $x = v/\|v\|_0$.) By property (ii) of norm we have that $\|v\|_1/\|v\|_2 = \|x\|_1/\|x\|_2$. This yields the required inequalities $\|v\|_1 \leq k_1\|v\|_2$, and $\|v\|_2 \leq k_2\|v\|_1$ with $k_1 = d$, and $k_2 = c^{-1}$. This completes the proof.

Chapter 4

EIGENVALUES AND EIGENVECTORS

We introduce the fundamental and important notions of eigenvalues and eigenvectors of a square matrix. We give the definitions, method of determination, and basic properties of eigenvalues and eigenvectors. We consider matrix polynomials, the Cayley-Hamilton theorem, and minimal polynomials of matrices. We then consider diagonalizable matrices, i.e. matrices similar to diagonal matrices, and determine exactly which matrices are diagonalizable. Finally we give a few examples of situations where the above notions are encountered.

4.1 **Eigenvalues and eigenvectors**

Let $M_n F$ be the set of all $n \times n$ matrices with entries in F, where $F = \mathbb{R}$ or $\mathbb{C}$. We will regard $M_n \mathbb{R}$ as a subset of $M_n \mathbb{C}$. For the consideration of eigenvalues it is convenient to think of a real-valued matrix as an element of $M_n \mathbb{C}$.

4.1.1 **Definition**

The complex number λ_0 is called an *eigenvalue* of the $n \times n$ matrix A provided that there exists a non-zero column vector v in $\mathbb{C}^n$ such that $Av = \lambda_0 v$.

Eigenvalues are also sometimes known as characteristic values, latent values or proper values. The name comes from the German word "eigenwert".

4.1.2 **Definition**

The non-zero vector v in (4.1.1) is called an *eigenvector* corresponding to the eigenvalue λ_0.

4.1.3 **Remark**

It is important that the vector v in (4.1.1) is non-zero. If we allowed v in (4.1.1) to be the zero vector then every complex number would be an eigenvalue of A !

Observe also that if v is an eigenvector for A corresponding to the eigenvalue λ_0 then so is αv for every non-zero $\alpha \in \mathbb{C}$.

4.2 **Determination of eigenvalues**

The equation $Av = \lambda_0 v$ in (4.1.1) can be rewritten in the form $(A - \lambda_0 I)v = 0$. This means that the non-zero vector v lies in the kernel of the linear map defined by the matrix $A - \lambda_0 I$.

4.2.1 **Definition**

The *characteristic polynomial* of the n×n matrix A is the polynomial in one variable λ given by $\det(A - \lambda I)$.

We write $p_A(\lambda)$ for this polynomial. It is a polynomial of degree n, i.e. the highest power of λ appearing is λ^n.

4.2.2 **Proposition**

The complex number λ_0 is an eigenvalue of the n×n matrix A if and only if λ_0 is a root of the characteristic polynomial $p_A(\lambda)$.
i.e. λ_0 is an eigenvalue of A if and only if $p_A(\lambda_0) = 0$.

Proof

If λ_0 is an eigenvalue of A then $(A - \lambda_0 I)v = 0$ for some non-zero vector v. Hence the linear map on $\mathbb{C}^n$ determined by $A - \lambda_0 I$ has non-trivial kernel. The matrix $A - \lambda_0 I$ must then be singular so that $\det(A - \lambda_0 I) = 0$.

Conversely if $p_A(\lambda_0) = 0$ then the matrix $A - \lambda_0 I$ is singular and so, by (2.6.5), there exists a non-zero vector v such that $(A - \lambda_0 I)v = 0$. Thus $Av = \lambda_0 v$ and λ_0 is an eigenvalue of A.

4.2.3 **Definition**

The *spectrum* of an n×n matrix A is the set of all the eigenvalues of A. We write $\sigma(A)$ for the spectrum of A.

4.2.4 **Exercise**

Show that $\sigma(A^t) = \sigma(A)$ for any n×n matrix A.

4.2.5 **Example**

$$A = \begin{pmatrix} 3 & 0 & 0 & 0 \\ 0 & 3 & 0 & 0 \\ 0 & 0 & 0 & -1 \\ 0 & 0 & 1 & 0 \end{pmatrix}$$

$$p_A(\lambda) = \det \begin{pmatrix} 3-\lambda & 0 & 0 & 0 \\ 0 & 3-\lambda & 0 & 0 \\ 0 & 0 & -\lambda & -1 \\ 0 & 0 & 1 & -\lambda \end{pmatrix} = (3 - \lambda)^2(\lambda^2 + 1)$$

$$= (3 - \lambda)^2(\lambda + i)(\lambda - i).$$

Hence $\sigma(A) = \{3, i, -i\}$.

Note that this example shows that, even if all of the entries of a matrix are real, the eigenvalues of the matrix need not all be real.

4.2.6 **Remarks**

We recall the Fundamental Theorem of Algebra which was first proved by Gauss in 1799. It says that any polynomial of degree n with real or complex coefficients factorizes over $\mathbb{C}$ into a product of linear factors.

Applying the theorem to the characteristic polynomial $p_A(\lambda)$ of an n×n matrix A we see that

$$p_A(\lambda) = (\lambda_1 - \lambda)(\lambda_2 - \lambda). \ . \ . \ . \ . (\lambda_n - \lambda)$$

for some complex numbers $\lambda_1, \lambda_2, . . ., \lambda_n$, these λ_i not necessarily being all different. Thus $p_A(\lambda)$ has n roots in $\mathbb{C}$, not necessarily all different.

Recall also that if all of the coefficients of the polynomial are real then any non-real roots occur in conjugate pairs, i.e. if λ_i is a root then so is $\bar{\lambda}_i$. Thus if $A \in M_n\mathbb{R}$ then any complex eigenvalues of A occur in conjugate pairs.

4.2.7 **Definition**

The *algebraic multiplicity* of the eigenvalue λ_0 of the n×n matrix A is its multiplicity as a root of the characteristic polynomial $p_A(\lambda)$.

In Example (4.2.5) the eigenvalue 3 has algebraic multiplicity two while the eigenvalues $\pm i$ each have algebraic multiplicity one.

4.2.8 **Definition**

The *eigenspace* for the eigenvalue λ_0 of the n×n matrix A is $\{v \in \mathbb{C}^n : Av = \lambda_0 v\}$. We will denote this eigenspace by W_{λ_0}.

Observe that W_{λ_0} is a subspace of $\mathbb{C}^n$ and it consists of all of the eigenvectors for λ_0 together with the zero vector. Note that W_{λ_0} is a non-zero subspace of $\mathbb{C}^n$ because there must exist a non-zero eigenvector.

4.2.9 **Definition**

The *geometric multiplicity* of the eigenvalue λ_0 of the n×n matrix A is the dimension of the eigenspace W_{λ_0}.

Its value will be a positive integer since $W_{\lambda_0} \neq 0$.

Determination of the eigenspace W_{λ_0} amounts to finding the solution set of the system of linear equations $Av = \lambda_0 v$. The dimension of this solution set will equal the geometric multiplicity. We illustrate this by using the matrix of Example (4.2.5).

The equation $Av = \lambda_0 v$ for this example yields

$$\begin{pmatrix} 3 & 0 & 0 & 0 \\ 0 & 3 & 0 & 0 \\ 0 & 0 & 0 & -1 \\ 0 & 0 & 1 & 0 \end{pmatrix} \begin{pmatrix} w \\ x \\ y \\ z \end{pmatrix} = \lambda_0 \begin{pmatrix} w \\ x \\ y \\ z \end{pmatrix}$$

on writing v as the column vector $\begin{pmatrix} w \\ x \\ y \\ z \end{pmatrix}$.

Solving the above system for the eigenvalue $\lambda_0 = 3$ yields a two-dimensional solution space so that $\lambda_0 = 3$ has geometric multiplicity two.

Specifically $W_3 = \left\{ \begin{pmatrix} w \\ x \\ 0 \\ 0 \end{pmatrix} : w \in \mathbb{C},\ x \in \mathbb{C} \right\}$

The eigenspaces for $\lambda_0 = \pm\, i$ are each one-dimensional and so these eigenvalues are each of geometric multiplicity one.

Specifically $W_i = \left\{ \begin{pmatrix} 0 \\ 0 \\ i\alpha \\ \alpha \end{pmatrix} : \alpha \in \mathbb{C} \right\}$ and $W_{-i} = \left\{ \begin{pmatrix} 0 \\ 0 \\ -i\alpha \\ \alpha \end{pmatrix} : \alpha \in \mathbb{C} \right\}$.

4.2.10 **Remark**

We will prove later (4.6.3) that the geometric multiplicity is always less than or equal to the algebraic multiplicity. Thus if λ_0 has algebraic multiplicity one then it must also have geometric multiplicity one.

4.3 Some properties of eigenvalues

We recall the notion of similar matrices which we encountered in (2.5.6).

4.3.1 **Definition**

Two nxn matrices A and B in M_nF, $F = \mathbb{R}$ or $\mathbb{C}$, are said to be *similar* if there exists an invertible nxn matrix $P \in M_nF$ such that $P^{-1}AP = B$.

4.3.2 **Proposition**

Similar matrices have the same eigenvalues.

Proof

Let $P^{-1}AP = B$ as in the above definition. We will show that A and B have the same characteristic polynomial and hence must have exactly the same eigenvalues.

$$\begin{aligned} p_B(\lambda) &= \det (B - \lambda I) \\ &= \det (P^{-1}AP - \lambda I) \\ &= \det (P^{-1}(A - \lambda I)P) \\ &= (\det P^{-1})(\det (A - \lambda I))(\det P) \\ &= \det (A - \lambda I) \\ &= p_A(\lambda). \end{aligned}$$

This completes the proof.

4.3.3 **Exercise**

If $P^{-1}AP = B$ and λ_0 is an eigenvalue of A, (and hence of B), show that the vector v is an eigenvector of B corresponding to the eigenvalue λ_0 if and only if Pv is an eigenvector of A corresponding to the same eigenvalue λ_0. Deduce that λ_0 has the same geometric multiplicity as an eigenvalue for both A and B.

4.3.4 **Comment**

Suppose that the nxn matrix A is one of the following three types;

(i) diagonal, i.e. $a_{ij} = 0$ for all $i \neq j$,

(ii) upper triangular, i.e. $a_{ij} = 0$ for all $i > j$,

(iii) lower triangular, i.e. $a_{ij} = 0$ for all $i < j$.

Then the eigenvalues of A are precisely the diagonal entries of A. This is because $p_A(\lambda) = \prod_{i=1}^{n} (a_{ii} - \lambda)$.

(The symbol Π here denotes product.)

We will see later, see (6.1.3), that any $A \in M_n\mathbb{C}$ is similar to an upper triangular matrix and also we will determine exactly which $A \in M_n\mathbb{C}$ are similar to a diagonal matrix. See (4.6) in this chapter.

Most computer algorithms for finding the eigenvalues of a matrix A are based on the idea of reducing A by similarity transformations to a matrix of type (i), (ii), or (iii). The eigenvalues of this simpler matrix can be read off immediately and, by (4.3.2), are the eigenvalues of A. The most famous such algorithm is known as the *QR-algorithm*. We will discuss this further in chapter 8.

4.3.5 **Proposition**

Let $\lambda_1, \lambda_2, \ldots, \lambda_n$ be the eigenvalues of the nxn matrix A, these λ_i not necessarily being distinct.

Then $\text{trace } A = \sum_{i=1}^{n} \lambda_i = \lambda_1 + \lambda_2 + \ldots + \lambda_n$

and $\det A = \prod_{i=1}^{n} \lambda_i = \lambda_1 \lambda_2 \cdots \lambda_n$.

Proof

We write $p_A(\lambda) = (\lambda_1 - \lambda)(\lambda_2 - \lambda) \ldots (\lambda_n - \lambda)$ since the λ_i are the roots of the characteristic polynomial. From this we see that the constant term of $p_A(\lambda)$ is $\lambda_1 \lambda_2 \cdots \lambda_n$ and the coefficient of λ^{n-1} in $p_A(\lambda)$ is $(-1)^{n-1}(\lambda_1 + \lambda_2 + \ldots + \lambda_n)$.

However we know also that $p_A(\lambda) = \det(A - \lambda I)$ and putting $\lambda = 0$ yields that the constant term in $p_A(\lambda)$ is det A.

Examining the expansion of $\det(A - \lambda I) = \det \begin{pmatrix} a_{11}-\lambda & a_{12} & a_{1n} \\ a_{21} & a_{22}-\lambda & a_{2n} \\ & & \\ a_{n1} & a_{n2} & a_{nn}-\lambda \end{pmatrix}$ we see that the coefficient of λ^{n-1} is $(-1)^{n-1}(a_{11} + a_{22} + \ldots + a_{nn})$ and since $\text{trace } A = \sum_{i=1}^{n} a_{ii}$ this completes the proof.

We have seen that the eigenvalues of a matrix need not be real, even when all of the entries of the matrix are real. However the following theorem shows that there are some important special cases when we can be sure that all of the eigenvalues are real.

4.3.6 **Proposition**

The eigenvalues of a real symmetric matrix or a complex hermitian matrix are real.

Proof

Let A be a complex hermitian matrix, i.e. $A = \bar{A}^t$. The argument we give is also valid in the real symmetric case where the conjugation operation on A will have no effect since all the entries of A are real.

Let λ be an eigenvalue of A with corresponding eigenvector $v \in \mathbb{C}^n$. At this stage we only know that $\lambda \in \mathbb{C}$.

Then $\bar{v}^t A v = \bar{v}^t \lambda v = \lambda \bar{v}^t v$, v being a column vector in $\mathbb{C}^n$. Now $\bar{v}^t v$ and $\bar{v}^t A v$ are each 1x1 matrices and in fact are real numbers because they are unchanged under the operation of conjugate transpose. (This follows by using the comment after (1.1.11) and the fact that $A = \bar{A}^t$.) Indeed $\bar{v}^t v$ is the square of the norm of v and so is non-zero. Hence λ must be a real number.

4.3.7 **Linear operators**

The notion of eigenvalue makes sense in the more general context of linear operators.

A linear operator on the vector space V over $\mathbb{R}$ or $\mathbb{C}$ is a linear map $T : V \longrightarrow V$. If V is finite-dimensional then T may be represented by a square matrix with respect to some chosen basis of V.

We say that λ_0 is an *eigenvalue* of T provided that there exists a non-zero element $v \in V$ such that $Tv = \lambda_0 v$. The vector v is called an *eigenvector* of T corresponding to the eigenvalue λ_0.

When V is of finite dimension the eigenvalues of T will be the eigenvalues of any matrix representing T. Recall from (2.5.6) that matrices representing T with respect to two different bases of V will be similar and hence, by (4.3.2), will have the same eigenvalues.

The study of operators on infinite-dimensional vector spaces lies outside our ambit and belongs rather to the subject known as functional analysis.

Problems 4A

1. Determine the eigenvalues and their algebraic and geometric multiplicities for each of the following matrices:

(i) $\begin{pmatrix} 2 & 6 & -1 \\ 0 & 1 & 3 \\ 0 & 3 & 1 \end{pmatrix}$ (ii) $\begin{pmatrix} 0 & 1 & 0 \\ 1 & 0 & 0 \\ 0 & 0 & 1 \end{pmatrix}$ (iii) $\begin{pmatrix} 1 & 2 & 0 \\ 2 & 1 & 2 \\ 0 & 2 & 1 \end{pmatrix}$

(iv) $\begin{pmatrix} i & i \\ 1 & 1 \end{pmatrix}$ (v) $\begin{pmatrix} i & i & 0 \\ 1 & 1 & 0 \\ 0 & 0 & 1+i \end{pmatrix}$ (vi) $\begin{pmatrix} 1 & 0 & 1 & 0 \\ 0 & 1 & 0 & 1 \\ 1 & 0 & 1 & 0 \\ 0 & 1 & 0 & 1 \end{pmatrix}$

2. A matrix A is said to be *idempotent* if $A^2 = A$. Show that the only possible eigenvalues of an idempotent matrix are zero and one.

3. A matrix A is said to be *nilpotent* if $A^k = 0$ for some natural number k. Show that zero is the only eigenvalue of a nilpotent matrix.

4. Show that the matrix A is invertible if and only if zero is not an eigenvalue of A.

Show that if A is invertible then $\sigma(A^{-1}) = \{ \lambda^{-1} : \lambda \in \sigma(A) \}$.

5. If A is a matrix of rank one show that trace A is an eigenvalue of A. Hence determine the eigenvalues of the nxn matrix $A = (a_{ij})$ with $a_{ij} = 1$ for

all i and j.

6. Let $A \in M_n\mathbb{C}$ and let v be a non-zero row vector in $\mathbb{C}^n$. If $vA = \lambda_0 v$ for some $\lambda_0 \in \mathbb{C}$ then λ_0 is called a *left eigenvalue* of A and v a *left eigenvector* corresponding to λ_0.

Let the eigenvalues and eigenvectors as we have defined them in (4.1) be called *right eigenvalues* and *right eigenvectors.*

Show that the row vector $v \in \mathbb{C}^n$ is a left eigenvector of A if and only if $\bar{v}^t$ is a right eigenvector for the matrix $\bar{A}^t$, and that the set of left eigenvalues of A coincides with the set of right eigenvalues of A.

7. Let $p(x) = a_0 + a_1x + a_2x^2 + \ldots + a_{n-1}x^{n-1} + x^n$.

The *companion matrix* of the polynomial p(x) is the n×n matrix

$$C = \begin{pmatrix} 0 & 1 & 0 & 0 & \ldots & 0 \\ 0 & 0 & 1 & 0 & \ldots & 0 \\ 0 & 0 & 0 & 1 & \ldots & 0 \\ & & & & & \\ 0 & 0 & 0 & 0 & \ldots & 1 \\ -a_0 & -a_1 & -a_2 & -a_3 & \ldots & -a_{n-1} \end{pmatrix}.$$

i.e. A has entries 1 on the superdiagonal, $-a_0, -a_1, \ldots, -a_{n-1}$ along the bottom row and zero everywhere else.

Show that the characteristic polynomial of C is $(-1)^n\, p(x)$.

(*Hint* - use induction on n after expanding det $(C - \lambda I)$ along the top row.)

8. Let A and B be n×n matrices. Show that AB and BA have exactly the same eigenvalues.

(*Hint* - show that the 2n×2n matrices $\begin{pmatrix} 0 & 0 \\ B & BA \end{pmatrix}$ and $\begin{pmatrix} AB & 0 \\ B & 0 \end{pmatrix}$ are similar via the matrix $\begin{pmatrix} I & A \\ 0 & I \end{pmatrix}$.)

9. Let λ_0 be an eigenvalue of the n×n matrix A. Show that any nonzero column of the adjugate of the matrix $A - \lambda_0 I$ is an eigenvector of A corresponding to the eigenvalue λ_0.

10. Let A and B are n×n matrices with real entries and suppose that they are similar over $\mathbb{C}$, i.e. there exists $P \in M_n\mathbb{C}$ such that $P^{-1}AP = B$. Show that A and B are similar over $\mathbb{R}$, i.e. there exists $Q \in M_n\mathbb{R}$ such that $Q^{-1}AQ = B$. (*Hint* - write $P = R + iS$ where $R \in M_n\mathbb{R}$, $S \in M_n\mathbb{R}$, show that $AR = RB$, $AS = SB$, and that if R is not invertible then $R + \alpha S$ is invertible for some real number α. Observe that $\det(R + \alpha S)$ is a non-constant polynomial in α.)

4.4 Eigenvalues and norms

4.4.1 Proposition

Let $\| \ \|$ be any operator norm on M_nF, $F = \mathbb{R}$ or $\mathbb{C}$, and let λ be any eigenvalue of the matrix $A \in M_nF$. Then $|\lambda| \le \|A\|$.

(i.e. any operator norm gives an upper bound on the size of the eigenvalues.)

Proof

Let v be an eigenvector for the eigenvalue λ of the matrix A. Then $Av = \lambda v$ implies that $\|Av\| = |\lambda| \ \|v\|$ using properties of norms. Thus

$$\frac{\|Av\|}{\|v\|} = |\lambda|$$

for this particular vector v and the definition of operator norm (3.3.8) shows that $|\lambda| \le \|A\|$.

4.4.2 Corollary

For each eigenvalue λ of the n×n matrix $A = (a_{ij})$ we have that

$$|\lambda| \le \max_i \left(\sum_{j=1}^{n} |a_{ij}| \right) \text{ and } |\lambda| \le \max_j \left(\sum_{i=1}^{n} |a_{ij}| \right).$$

Proof

Immediate from (4.4.1) and (3.4.2).

4.4.3 **Example**

$$A = \begin{pmatrix} 0 & 0.1 & -0.12 & 0.2 & 0.3 \\ 0.2 & 0 & 0.2 & 0 & 0.4 \\ 0.3 & -0.3 & 0 & 0 & 0.3 \\ 0.2 & 0.15 & -0.3 & 0 & 0.3 \\ -0.1 & 0.2 & 0.2 & 0.2 & 0 \end{pmatrix}$$

The absolute row sums for A are 0.72, 0.8, 0.9, 0.95, and 0.7. Thus $|\lambda| \leq 0.95$ for each eigenvalue λ of A.

Problems 4B

1. An n×n matrix A is called *stochastic* if all of the entries of A are non-negative and the sum of the entries in each column is one. Show that $|\lambda| \leq 1$ for each eigenvalue λ of A.

2. Let $p(x) = a_0 + a_1x + a_2x^2 + \ldots + a_{n-1}x^{n-1} + x^n$ be a polynomial with coefficients $a_i \in \mathbb{C}$, $i = 1, 2, \ldots, n$.

Show that all of the roots of p(x) lie in a disc in $\mathbb{C}$ with centre zero, radius $r = \max\{|a_0|, 1 + |a_1|, 1 + |a_2|, \ldots, 1 + |a_{n-1}|\}$. (Use (4.4.2) and problem 7 of Problems 4A.)

Deduce that all the roots of $x^9 + x^6 + 2x^4 - 2x + 1$ lie in the disc $|x| \leq 3$.

4.5 **Polynomials in a matrix**

Let $p(\lambda) = a_0 + a_1\lambda + a_2\lambda^2 + \ldots\ldots + a_m\lambda^m$ be a polynomial in one variable λ and with coefficients $a_i \in \mathbb{C}$ for each $i = 1, 2, \ldots, m$.

Let A be an n×n matrix. Then the powers A^k for $k = 1,2,3,\ldots$ make sense and are each n×n matrices. Furthermore, via the following definition, we can give meaning to the notion of the polynomial p being evaluated at the matrix A.

4.5.1 **Definition**

$p(A) = a_0I + a_1A + a_2A^2 + \ldots\ldots + a_mA^m$ where I denotes the identity nxn matrix.

Thus p(A) is an nxn matrix.

If p(A) = 0, the zero matrix, we say that A is *annihilated* by the polynomial p.

Recall from (4.2.1) that the characteristic polynomial $p_A(\lambda)$ of the nxn matrix A is defined by $p(\lambda) = \det(A - \lambda I)$.

4.5.2 **Cayley-Hamilton theorem**

Any nxn matrix A is annihilated by its characteristic polynomial, i.e. $p_A(A) = 0$ for any $A \in M_n\mathbb{C}$.

Proof

Recall from (1.4.7) that for any nxn matrix C we have an equation C (adj C) = (det C) I where adj C is the adjugate of C.

Letting $C = A - \lambda I$ we have $(A - \lambda I)\ \mathrm{adj}\ (A - \lambda I) = p_A(\lambda)\ I$. The entries of adj $(A - \lambda I)$ will be polynomials in λ of degree at most n - 1. Thus we may write $\mathrm{adj}\ (A - \lambda I) = \sum_{i=0}^{n-1} B_i\lambda^i$ where each $B_i \in M_n\mathbb{C}$.

If we write $p_A(\lambda) = \sum_{i=0}^{n} \alpha_i\lambda^i$ where each $\alpha_i \in \mathbb{C}$ then the above equation becomes $(A - \lambda I)\left(\sum_{i=0}^{n-1} B_i\lambda^i\right) = \left(\sum_{i=0}^{n} \alpha_i\lambda^i\right)$.

Multiplying out the left-hand side of this equation and then equating powers of λ^i for i = 0,1,2, ... ,n yields the following set of n + 1 matrix equations ;

$$AB_0 = \alpha_0 I$$

$$AB_1 - B_0 = \alpha_1 I$$

$$AB_2 - B_1 = \alpha_2 I$$

$$AB_{n-1} - B_{n-2} = \alpha_{n-1} I$$

$$B_{n-1} = \alpha_n I$$

Multiplying the second equation by A, the third by A^2, and so on until the last one by A^n, and then adding all the equations yields that $0 = \sum_{i=0}^{n} \alpha_i A^i$, i.e. $p_A(A) = 0$, and the proof is complete.

4.5.3 **Exercise**

Verify the Cayley-Hamilton theorem for the matrix $A = \begin{pmatrix} 7 & -1 \\ 2 & 4 \end{pmatrix}$.

4.5.4 **Definition**

The *minimal polynomial* of the nxn matrix A is the monic polynomial $m(\lambda)$ of least degree such that $m(A) = 0$.

(A *monic* polynomial is one in which the coefficient of the highest power of λ is one.)

Note that $m(\lambda)$ is unique since if we have polynomials $m_1(\lambda)$ and $m_2(\lambda)$ each satisfying definition (4.5.4) then $m_1(\lambda) - m_2(\lambda)$ is a polynomial of lower degree which annihilates A. Dividing by the leading coefficient of $m_1(\lambda) - m_2(\lambda)$ we will obtain a monic polynomial of lower degree which annihilates A.

Note that the characteristic polynomial $p_A(\lambda) = \det(A - \lambda I)$ of the nxn matrix A is monic if n is even while $-p_A(\lambda)$ is monic if n is odd. Since p_A

annihilates A it follows that the minimal polynomial has degree less than or equal to that of the characteristic polynomial.

4.5.5 **Proposition**

Let $p_A(\lambda)$ and $m_A(\lambda)$ denote the characteristic and minimal polynomials respectively of the matrix $A \in M_n\mathbb{C}$. Then $m_A(\lambda)$ is a divisor of $p_A(\lambda)$. Furthermore $m_A(\lambda)$ and $p_A(\lambda)$ have exactly the same roots in $\mathbb{C}$ although not necessarily occurring with the same algebraic multiplicities.

Proof

Since the degree of $m_A(\lambda)$ is less than or equal to the degree of $p_A(\lambda)$ we may perform long division and find that for suitable polynomials $q(\lambda)$ and $r(\lambda)$, with $r(\lambda)$ of lower degree than $m_A(\lambda)$,

$$p_A(\lambda) = q(\lambda)\, m_A(\lambda) + r(\lambda)$$

We know that $p_A(A) = 0$ and $m_A(A) = 0$ and so the above equation gives $r(A) = 0$. This would contradict the minimality of $m_A(\lambda)$ unless $r(\lambda)$ is identically zero.

This proves that $m_A(\lambda)$ is a divisor of $p_A(\lambda)$.

To prove the final assertion of the proposition we must show that λ_0 is an eigenvalue of A if and only if $m_A(\lambda_0) = 0$.

Suppose first that λ_0 is an eigenvalue of A, so that $Av = \lambda_0 v$ for some $v \in \mathbb{C}^n$, $v \neq 0$. Then $A^k v = \lambda_0^k v$ for $k = 1.2. \ldots$, and this implies that $p(A)v = p(\lambda_0)v$ for any polynomial p. If we take $p = m_A$ we find that $m_A(A)v = m_A(\lambda_0)v$. Since $m_A(A) = 0$ and $v \neq 0$ we have that $m_A(\lambda_0) = 0$.

Conversely if $m_A(\lambda_0) = 0$ then $m_A(\lambda) = (\lambda - \lambda_0)\, t(\lambda)$ for some polynomial $t(\lambda)$ of degree $m - 1$ where m is the degree of $m_A(\lambda)$. Now $t(A) \neq 0$ by the minimality of $m_A(\lambda)$ and so $t(A)(v) \neq 0$ for some $v \in \mathbb{C}^n$. Putting $\lambda = A$ in the above equation we find that $m_A(A) = (A - \lambda_0 I)\, t(A)$, i.e. $(A - \lambda_0 I)\, t(A) = 0$. Thus λ_0 is an eigenvalue of A with the non-zero vector $t(A)v$ as corresponding eigenvector. This completes the proof.

4.5.6 Example

$$A = \begin{pmatrix} 2 & 1 & 0 & 0 \\ 0 & 2 & 0 & 0 \\ 0 & 0 & 2 & 0 \\ 0 & 0 & 0 & 5 \end{pmatrix}$$

A is upper triangular so that $p_A(\lambda) = (2 - \lambda)^3(5 - \lambda)$. It follows from (4.5.5) that there are three possibilities for the minimal polynomial $m_A(\lambda)$, namely $(\lambda - 2)(\lambda - 5)$, $(\lambda - 2)^2(\lambda - 5)$ and $(\lambda - 2)^3(\lambda - 5)$. We proceed by putting $\lambda = A$ in each of these polynomials, starting with the one of lowest degree. It is an easy exercise to check that $(A - 2I)(A - 5I) \neq 0$ whereas $(A - 2I)^2(A - 5I) = 0$. This shows that $m_A(\lambda) = (\lambda - 2)^2(\lambda - 5)$.

4.5.7 Comment

In principle the minimal polynomial can be calculated as follows ;

Determine the characteristic polynomial, factorize it completely, and list all of the possibilities for the minimal polynomial in increasing order of degree. Evaluate each possibility at A, starting at the one of lowest degree, until a polynomial which annihilates A is obtained.

For small matrices this method is fine but for ones whose eigenvalues have large algebraic multiplicities this is not a good algorithmic method. Unfortunately there seems to be no especially quick general method of calculating minimal polynomials.

4.5.8 Lemma

Similar matrices have the same minimal polynomial.

Proof

If A and B are similar then $B = S^{-1}AS$ for some invertible matrix S. Let p be any polynomial. Since $B^k = S^{-1}A^kS$ for all natural numbers k it follows that $p(B) = S^{-1}p(A)S$. Thus $p(B) = 0$ if and only if $p(A) = 0$ and in particular A and B must have the same minimal polynomial.

Problems 4C

1. Let A be a 2x2 matrix such that $A^2 = I$ but $A \neq \pm I$. Deduce that trace A = 0 and det A = -1.

(*Hint* - show that $p_A(\lambda) = \lambda^2 - (\text{trace } A)\lambda + \det A$.)

2. Use the Cayley-Hamilton theorem to show that if A is a non-singular nxn matrix then there is a polynomial p of degree less than or equal to n - 1 such that $A^{-1} = p(A)$.

3. Find the minimal polynomial of each of the following matrices:

(i) $\begin{pmatrix} 2 & 0 & 0 \\ 0 & 2 & 2 \\ 0 & 0 & 2 \end{pmatrix}$ (ii) $\begin{pmatrix} 0 & 0 & 3 \\ 0 & 0 & 4 \\ 0 & 0 & 0 \end{pmatrix}$ (iii) $\begin{pmatrix} 0 & 1 & 0 & 1 \\ 1 & 0 & 1 & 0 \\ 0 & 1 & 0 & 1 \\ 1 & 0 & 1 & 0 \end{pmatrix}$

(iv) $\begin{pmatrix} 2 & 1 & 0 & 0 \\ 0 & 2 & 0 & 0 \\ 0 & 0 & 2 & 0 \\ 0 & 0 & 0 & 5 \end{pmatrix}$ (v) $\begin{pmatrix} 0 & 0 & 0 & 0 \\ 1 & 1 & 1 & 1 \\ 1 & 1 & 1 & 1 \\ 1 & 1 & 1 & 1 \end{pmatrix}$

4. Give an example of a 3x3 matrix whose minimal polynomial is λ^2 and of a 4x4 matrix whose minimal polynomial is λ^3.

5. Let $A = \begin{pmatrix} 0 & 0 & c \\ 1 & 0 & b \\ 0 & 1 & a \end{pmatrix}$ where a,b, and c are real numbers. Show that the minimal polynomial of A is $\lambda^3 - a\lambda^2 - b\lambda - c$.

6. Let $p(x) = a_0 + a_1x + a_2x^2 + \ldots + a_{n-1}x^{n-1} + x^n$ and let C be its companion matrix as defined in problem 7 of Problems 4A.

$$C = \begin{pmatrix} 0 & 1 & 0 & 0 & \ldots & 0 \\ 0 & 0 & 1 & 0 & \ldots & 0 \\ 0 & 0 & 0 & 1 & \ldots & 0 \\ & & & & & \\ 0 & 0 & 0 & 0 & \ldots & 1 \\ -a_0 & -a_1 & -a_2 & -a_3 & \ldots & -a_{n-1} \end{pmatrix}.$$

Show that the minimal polynomial of C is equal to p(x).

4.6 Diagonalizable matrices

4.6.1 Definition

Let $A \in M_n F$ where $F = \mathbb{R}$ or $\mathbb{C}$. The matrix A is said to be *diagonalizable over F* if there exist matrices $S \in M_n F$, $D \in M_n F$ with S invertible, D diagonal, and $S^{-1}AS = D$.

(i.e. A is diagonalizable over F if and only if A is similar to a diagonal matrix in $M_n F$.)

We will obtain necessary and sufficient conditions for matrices to be diagonalizable over $\mathbb{R}$ or $\mathbb{C}$. First we need some background material.

4.6.2 Proposition

Let $A \in M_n F$. Let $\lambda_1, \lambda_2, \ldots, \lambda_k$ be any set of eigenvalues of A, the λ_i being all different, and let $v_1, v_2, \ldots, v_k$ be a set of eigenvectors with v_i corresponding to λ_i for each i. Then $\{v_1, v_2, \ldots, v_k\}$ is a linearly independent set.

Proof

Assume that $\{v_1, v_2, \ldots, v_k\}$ is linearly dependent. Then the zero vector is expressible as a non-trivial linear combination of $\{v_1, v_2, \ldots, v_k\}$.

We write the zero vector as a non-trivial linear combination of $\{v_1, v_2, \ldots, v_k\}$ of minimal length,

$$0 = \alpha_1 v_{i_1} + \alpha_2 v_{i_2} + \ldots\ldots + \alpha_s v_{i_s}$$

where each $\alpha_j \neq 0$, each $v_{i_j} \in \{v_1, v_2, \ldots, v_k\}$, and s is minimal, i.e. no other expression of the zero vector as a linear combination of $\{v_1, v_2, \ldots, v_k\}$ has less than s non-zero coefficients.

Applying A to this minimal length equation and using the fact that the v_i are eigenvectors yields a new equation

$$0 = \alpha_1\lambda_{i_1}v_{i_1} + \alpha_2\lambda_{i_2}v_{i_2} + \ldots + \alpha_s\lambda_{i_s}v_{i_s}$$

Assume that $\lambda_{i_1} \neq 0$. (If $\lambda_{i_1} = 0$ then we do the same argument using λ_{i_2} instead!) Multiplying our first equation by λ_{i_1} and subtracting from the second equation yields a new expression for the zero vector in terms of $\{v_1, v_2, \ldots, v_k\}$ and this expression has length less than s. This is a contradiction and so the proof is complete.

4.6.3 **Lemma**

Let λ_0 be any eigenvalue of the $n \times n$ matrix A. Then the geometric multiplicity of λ_0 is less than or equal to the algebraic multiplicity.

Proof

Let W_0 be the eigenspace for the eigenvalue λ_0 and suppose W_0 has dimension k. We must show that λ_0 occurs at least k times as a root of the characteristic polynomial $p_A(\lambda)$.

Let $f : F^n \longrightarrow F^n$, $f(v) = Av$ for each $v \in F^n$ be the linear map which is represented with respect to the standard basis of F^n by the matrix A. Choose some basis for W_0 and extend it to a basis of F^n. (This is always possible by (2.2.13) !) Then the matrix B of the map f with respect to this new basis is of the form

$$B = \begin{pmatrix} \lambda_0 & 0 & 0 & * & * & * \\ 0 & \lambda_0 & 0 & * & * & * \\ & & & * & * & * \\ 0 & 0 & \lambda_0 & * & * & * \\ 0 & 0 & 0 & * & * & * \\ & & & * & * & * \\ & & & * & * & * \\ 0 & 0 & 0 & * & * & * \end{pmatrix}$$

i.e. B has a $k \times k$ block $\lambda_0 I$ in its top left-hand corner and all other entries in the first k columns are zero.

From the shape of B its characteristic polynomial is given by $p_B(\lambda) = (\lambda_0 - \lambda)^k q(\lambda)$ for some polynomial $q(\lambda)$. Now, by (2.5.6), A and B are similar and so, by (4.3.2), they have the same characteristic polynomial and hence the same eigenvalues occurring with the same algebraic multiplicities. Thus the algebraic multiplicity of λ_0 is at least k and the proof is complete.

4.6.4 **Corollary**

If λ_0 has algebraic multiplicity one then it also has geometric multiplicity one.

Proof

The geometric multiplicity is at least one because, by the definition of eigenvalue, there must exist a non-zero element in the eigenspace.

4.6.5 **Theorem**

Let $A \in M_nF$ where $F = \mathbb{R}$ or $\mathbb{C}$. The matrix A is diagonalizable over F if and only if F^n has a basis consisting entirely of eigenvectors of A.

Proof

If $\{v_1, v_2, ... ,v_n\}$ is a basis of F^n with each v_i an eigenvector of A then $Av_i = \lambda_i v_i$ where $\lambda_1, \lambda_2, . . . ,\lambda_n$ are the eigenvalues, not necessarily all different. Then A is similar to the diagonal matrix with $\lambda_1, \lambda_2, . . ,\lambda_n$ on the diagonal. The matrix P having $v_1, v_2, . . ,v_n$ as its columns provides the similarity transformation. (This matrix P is the matrix of change of basis from the standard basis to the basis of eigenvectors.)

Conversely if there exists an invertible matrix P and a diagonal matrix D such that $P^{-1}AP = D$ then the columns of P are necessarily eigenvectors corresponding to eigenvalues which are the diagonal entries of D. (Consider the equation AP = PD and equate columns!) Since P is invertible the columns of P must form a linearly independent set and hence a basis for F^n.

4.6.6 **Special case**

If $A \in M_nF$ has n distinct eigenvalues in F then A is diagonalizable over F.

Proof

By (4.6.2) the n eigenvectors corresponding to the n distinct eigenvalues of A will form a linearly independent set and hence they are a basis of F^n.

4.6.7 **Comment**

Let $A \in M_n\mathbb{R}$. If at least one of the eigenvalues of A is not real then A is not diagonalizable over $\mathbb{R}$. There is no chance of getting enough eigenvectors of A to form a basis of $\mathbb{R}^n$!

4.6.8 **Summary**

(i) Let $A \in M_n\mathbb{C}$. Then A is diagonalizable over $\mathbb{C}$ if and only if for each eigenvalue of A the algebraic and geometric multiplicities coincide.

(ii) Let $A \in M_n\mathbb{R}$. Then A is diagonalizable over $\mathbb{R}$ if and only if all of the eigenvalues of A are real and the algebraic and geometric multiplicities coincide for each eigenvalue.

4.6.9 **Exercise**

Show that $A = \begin{pmatrix} 3 & 2 & 1 & 0 \\ 0 & 1 & 3 & 0 \\ 0 & 0 & 2 & 1 \\ 0 & 0 & 1 & 2 \end{pmatrix}$ is not diagonalizable over $\mathbb{C}$.

4.6.10 **Remark on terminology**

A matrix which is not diagonalizable over $\mathbb{C}$ is sometimes called a *defective* matrix.

Problems 4D

1. Show that $A = \begin{pmatrix} 2 & -1 & 0 \\ -1 & 2 & 0 \\ 2 & 2 & 3 \end{pmatrix}$ is diagonalizable over $\mathbb{R}$ and find a matrix P such that $P^{-1}AP$ is diagonal.

2. For each of the following matrices A determine whether or not A is diagonalizable over $\mathbb{C}$ and find P such that $P^{-1}AP$ is diagonal when A is diagonalizable.

(i) $\begin{pmatrix} 2 & 3 \\ 6 & -1 \end{pmatrix}$ (ii) $\begin{pmatrix} 2 & 1 & 0 & 0 \\ 0 & 2 & 0 & 0 \\ 0 & 0 & 2 & 0 \\ 0 & 0 & 0 & 5 \end{pmatrix}$ (iii) $\begin{pmatrix} 1 & -10 & 0 \\ -1 & 3 & 1 \\ -1 & 0 & 4 \end{pmatrix}$

3. Let $A = \begin{pmatrix} 7 & -5 & 15 \\ 6 & -4 & 15 \\ 0 & 0 & 1 \end{pmatrix}$. Find a matrix P such that $P^{-1}AP$ is diagonal and hence calculate A^6.

4. Let $A = \begin{pmatrix} -2 & 6 \\ 6 & 3 \end{pmatrix}$ and $B = \begin{pmatrix} 4 & 1 \\ 1 & 4 \end{pmatrix}$.

Show that A and B are each diagonalizable over $\mathbb{R}$.

Find matrices P and Q such that $P^{-1}AP$ and $Q^{-1}BQ$ are each diagonal.

Let $C = \begin{pmatrix} -2 & 6 & 0 & 0 \\ 6 & 3 & 0 & 0 \\ 0 & 0 & 4 & 1 \\ 0 & 0 & 1 & 4 \end{pmatrix}$.

Use the earlier part of this problem to find a 4x4 matrix R such that $R^{-1}CR$ is diagonal.

4.7 Some places where eigenvalues and eigenvectors are used

(a) Calculus of several variables

In (1.7)(c) we met the Hessian matrix of second order partial derivatives of a function of several real variables. This is a symmetric matrix and so, by (4.3.6), all of its eigenvalues are real. The eigenvalues of the Hessian matrix are useful in the classification of the critical points of the function. Specifically a critical point is said to be *non-degenerate* if all of the eigenvalues of the Hessian matrix at that point are non-zero, and a non-degenerate critical point is said to be *of type (p,q)* if there are exactly p positive and q negative eigenvalues.

(b) **Probability**

In (1.7)(g) we met Markov processes which are described by a stochastic matrix called the transition matrix of the process. It turns out that such a matrix always has the real number one occurring as an eigenvalue and it occurs with algebraic multiplicity one. The eigenvector corresponding to this eigenvalue will describe the long-term state of the process. All of this will be dealt with in detail in chapter 5.

(c) **Systems of differential equations and control theory**

In (1.7)(e) we met the equations governing feedback control systems:

$$\dot{x} = Ax + By, \; y = Cx$$

If the $n \times n$ matrix A happens to be diagonalizable so that $P^{-1}AP = D$ where D is diagonal and P is invertible then the substitution $x = Pz$, where $z = (z_i)$ is a column vector of new variables, yields new equations ;

$$\dot{z} = Dz + P^{-1}By, \quad y = CPz$$

This is a considerable simplification because the new equation $\dot{z} = Dz + P^{-1}By$ gives a set of equations, the first of which involves only the variable z_1, the second involves only the variable z_2, etc.

Generally, a system of linear differential equations $\dot{x} = Ax$ is easily solved when A is a diagonalizable matrix. The technique is to perform a similarity transformation changing A to a diagonal matrix, solve the problem for the diagonal case, and do the reverse similarity transformation to obtain the solution of the original problem. We will see all this in detail in chapter 5.

(d) **Geometry**

The *general equation of second degree in* $\mathbb{R}^2$ is of the form

$$ax^2 + 2bxy + cy^2 + dx + ey + f = 0$$

where a, b, c, d, e, and f are constants, at least one of a, b, and c being non-zero.

The set of points $(x,y) \in \mathbb{R}^2$ satisfying such an equation will in general be some kind of curve in the plane, the specific nature of the curve depending on the constants a,b,c,d,e, and f.

The matrix $\begin{pmatrix} a & b \\ b & c \end{pmatrix}$ turns out always to be diagonalizable via a suitable matrix P which has the property that $P^{-1} = P^t$. We thus obtain $P^tAP = \begin{pmatrix} \lambda_1 & 0 \\ 0 & \lambda_2 \end{pmatrix}$ where λ_1, λ_2 are the eigenvalues of A.

A change of variables given by the substitution $\begin{pmatrix} x \\ y \end{pmatrix} = P \begin{pmatrix} u \\ v \end{pmatrix}$ leads to the second degree equation becoming

$$\lambda_1 u^2 + \lambda_2 v^2 + (d\ e)\, P \begin{pmatrix} u \\ v \end{pmatrix} + f = 0$$

Completing the square then yields an equation of the form

$$\lambda_1 X^2 + \lambda_2 Y^2 + k = 0$$

where X and Y are two new variables and k is a constant. (This is provided that λ_1 and λ_2 are both non-zero.)

If $\lambda_1\lambda_2 > 0$ then the curve will be an ellipse, a single point or no real points.

If $\lambda_1\lambda_2 < 0$ then the curve will be a hyperbola, or a pair of straight lines.

If one of λ_1 and λ_2 is zero then, after completing the square, the equation will become that representing a parabola.

Geometrically the diagonalization above amounts to rotation of the co-ordinate axes while completion of the square amounts to translation of the axes.

The *general equation of second degree in* $\mathbb{R}^3$ is of the form

$$ax^2 + by^2 + cz^2 + 2hxy + 2gxz + 2fyz + rx + sy + tz + u = 0$$

where a, b, c, h, g, f, r ,s, t, and u are constants and at least one of a, b, c, h, g, f is non-zero.

The matrix $\begin{pmatrix} a & h & g \\ h & b & f \\ g & f & c \end{pmatrix}$ turns out to be diagonalizable via a suitable matrix P for which $P^{-1} = P^t$ and the diagonal matrix obtained has the eigenvalues λ_1, λ_2, and λ_3 of A on the diagonal. As in the previous case in $\mathbb{R}^2$ we can, after suitable substitution and completion of the square, reduce the equation to a standard form.

The set of points in $\mathbb{R}^3$ satisfying the equation will in general be some kind of *quadric surface*, e.g. ellipsoid, hyperboloid, elliptic paraboloid etc., the specific nature of the surface depending on the values of λ_1, λ_2, and λ_3.

Chapter 5

THE JORDAN CANONICAL FORM AND APPLICATIONS

We have seen in chapter 4 that not every square matrix is diagonalizable over $\mathbb{C}$. However every square matrix is similar to a matrix which is "nearly diagonal" in a certain sense. This nearly diagonal matrix is known as the *Jordan canonical form* and is important both for theoretical purposes and practical applications.

In this chapter we begin by discussing nilpotent matrices and then proceed to the Jordan canonical form of a matrix and ways of determining this form. We then give some applications, in particular to solving systems of linear differential equations, to Markov processes, and to certain matrix equations.

5.1 **Nilpotent matrices**

5.1.1 **Definition**

An $n \times n$ matrix A is *nilpotent* if $A^k = 0$ for some natural number k.

The least such integer k for which $A^k = 0$ is called the *index of nilpotency* of A.

5.1.2 **Example**

$A = \begin{pmatrix} 0 & 1 \\ 0 & 0 \end{pmatrix}$ is nilpotent of index two.

5.1.3 **Example**

$$A = \begin{pmatrix} 0 & 1 & 0 \\ 0 & 0 & 1 \\ 0 & 0 & 0 \end{pmatrix}$$ is nilpotent of index three.

5.1.4 **Example**

$$A = \begin{pmatrix} 0 & a & b \\ 0 & 0 & c \\ 0 & 0 & 0 \end{pmatrix}$$ is nilpotent of index three for any non-zero values of a, b, and c.

5.1.5 **Example**

Let A be the nxn matrix with entries one along the superdiagonal, and zero everywhere else. Then A is nilpotent of index n.

5.1.6 **Definition**

A linear operator $f : V \longrightarrow V$ on a vector space V is *nilpotent* if $f^k = 0$ for some natural number k.

(Here f^k means the n-fold composite of the map f.)

It is clear that f is nilpotent if and only if any matrix representing f is nilpotent.

5.1.7 **Example**

Let V be the vector space of all polynomials of degree less than or equal to n. The space V has dimension n + 1 with $\{1, x, x^2, \ldots, x^n\}$ being a natural basis.

Let $f : V \longrightarrow V$ be the differentiation operator on V, i.e. define $f(p(x)) = p'(x)$, the derivative of p(x).

Then f is nilpotent of degree n + 1, the matrix of f with respect to the above basis being

$$\begin{pmatrix} 0 & 1 & 0 & 0 & & 0 \\ 0 & 0 & 2 & 0 & & 0 \\ 0 & 0 & 0 & 3 & & 0 \\ & & & & & \\ 0 & 0 & 0 & 0 & & n \\ 0 & 0 & 0 & 0 & & 0 \end{pmatrix}.$$

5.1 8 **Remark**

We will see shortly that any nilpotent matrix is similar to a block diagonal matrix with blocks like the matrix in (5.1.5).

5.2 **Jordan canonical form**

From (4.6.9) we know that not every matrix is diagonalizable over $\mathbb{C}$. We will find soon that every matrix is similar to a matrix of the form D + N where D is diagonal and N is nilpotent.

5.2.1 **Definition**

A *kxk Jordan block* $J(\lambda)$ corresponding to an element $\lambda \in \mathbb{C}$ is a matrix of the form

$$\begin{pmatrix} \lambda & 1 & 0 & 0 & & 0 \\ 0 & \lambda & 1 & 0 & & 0 \\ 0 & 0 & \lambda & 1 & & 0 \\ & & & & & \\ 0 & 0 & 0 & 0 & \lambda & 1 \\ 0 & 0 & 0 & 0 & 0 & \lambda \end{pmatrix}.$$

i.e. $J(\lambda)$ has entries λ on the diagonal, one on the superdiagonal, and zero elsewhere.

For k = 1 we define $J(\lambda) = (\lambda)$, the 1x1 matrix with entry λ.

Observe that $J(\lambda) = \lambda I + N$ where N is nilpotent.

5.2.2 **Jordan form theorem**

Any matrix $A \in M_n\mathbb{C}$ is similar to a matrix J of the following form:

$$J = \begin{pmatrix} J_1(\lambda_1) & 0 & 0 & & 0 \\ 0 & J_2(\lambda_2) & 0 & & 0 \\ 0 & 0 & J_3(\lambda_3) & & 0 \\ & & & & \\ 0 & 0 & 0 & & J_r(\lambda_r) \end{pmatrix}$$

where each $J_i(\lambda_i)$ is a Jordan block corresponding to an eigenvalue λ_i of A.

Each block $J_i(\lambda_i)$ is an $n_i x n_i$ matrix for some integer n_i.

The same eigenvalue λ_i may occur in more than one block, i.e. the λ_i need not all be different.

The matrix J is unique up to a permutation of the blocks $J_i(\lambda_i)$ and is known as the *Jordan canonical form* of A. (We will often refer to it simply as the *Jordan form* of A.)

The following information is useful in determining J ;

(i) The sum of the sizes of the blocks involving a particular eigenvalue of A equals the algebraic multiplicity of that eigenvalue.

(ii) The number of blocks involving a particular eigenvalue of A equals the geometric multiplicity of that eigenvalue.

(iii) The largest block involving a particular eigenvalue of A equals the multiplicity of that eigenvalue as a root of the minimal polynomial of A.

Proof

See the appendix to this chapter.

5.2.3 **Remark**

From (i) and (ii) of (5.2.2) we see that the Jordan form of the matrix A consists entirely of 1x1 blocks if and only if the algebraic and geometric multiplicities coincide for each eigenvalue of A. This is of course precisely the criterion for diagonalizability which we have already seen in (4.6.8).

Note also that the uniqueness of the Jordan form means that two similar matrices will have the same Jordan form, at least up to permutation of the blocks.

5.2.4 **Example**

$$A = \begin{pmatrix} 0 & 1 & 2 \\ 0 & 0 & 1 \\ 0 & 0 & 0 \end{pmatrix}.$$

The matrix A has only the eigenvalue $\lambda = 0$ which has algebraic multiplicity three and geometric multiplicity one. The eigenspace for $\lambda = 0$ is easily seen to be $\{ \alpha(1,0,0) : \alpha \in \mathbb{C} \}$.

The Jordan form $J = \begin{pmatrix} 0 & 1 & 0 \\ 0 & 0 & 1 \\ 0 & 0 & 0 \end{pmatrix}$.

5.2.5 **Example**

$$A = \begin{pmatrix} -2 & -1 & -3 \\ 4 & 3 & 3 \\ -2 & 1 & -1 \end{pmatrix}.$$

It is straightforward to calculate that A has characteristic polynomial $(2-\lambda)^2(-4-\lambda)$ and that $\lambda = 2$ occurs with geometric multiplicity one. Hence the Jordan form J of A is $\begin{pmatrix} -4 & 0 & 0 \\ 0 & 2 & 1 \\ 0 & 0 & 2 \end{pmatrix}$.

5.2.6 **Example**

Let A be a 7x7 matrix whose characteristic polynomial is $(2-\lambda)^4(3-\lambda)^3$ and whose minimal polynomial is $(2-\lambda)^2(3-\lambda)^2$.

Then corresponding to the eigenvalue $\lambda = 3$ there must be one 2x2 Jordan block and one 1x1 Jordan block. Corresponding to $\lambda = 2$ there must be at least one 2x2 Jordan block. Hence there must be either two or three Jordan blocks for $\lambda = 2$, according as to whether the geometric multiplicity of $\lambda = 2$ is two or three.

The following are the two possibilities for the Jordan form of A,

$$\begin{pmatrix} 2 & 1 & 0 & 0 & & & 0 \\ 0 & 2 & 0 & 0 & & & 0 \\ 0 & 0 & 2 & 1 & & & \\ 0 & 0 & 0 & 2 & & & \\ & & & & 3 & 1 & 0 \\ & & & & 0 & 3 & 0 \\ 0 & 0 & & & 0 & 0 & 3 \end{pmatrix} \text{ or } \begin{pmatrix} 2 & 1 & 0 & 0 & & & 0 \\ 0 & 2 & 0 & 0 & & & 0 \\ 0 & 0 & 2 & 0 & & & \\ 0 & 0 & 0 & 2 & & & \\ & & & & 3 & 1 & 0 \\ & & & & 0 & 3 & 0 \\ 0 & 0 & & & 0 & 0 & 3 \end{pmatrix},$$

depending on the geometric multiplicity of the eigenvalue $\lambda = 2$.

5.2.7 **Example**

$$A = \begin{pmatrix} 0 & 1 & 1 & 0 & 1 \\ 0 & 0 & 1 & 1 & 1 \\ 0 & 0 & 0 & 0 & 0 \\ 0 & 0 & 0 & 0 & 0 \\ 0 & 0 & 0 & 0 & 0 \end{pmatrix}$$

It is easy to see that A has λ^5 as characteristic polynomial and λ^3 as minimal polynomial. The geometric multiplicity of the eigenvalue $\lambda = 0$ is easily checked to be three. Hence there must be one 3x3 Jordan block and two 1x1 blocks, i.e. the Jordan form of A is as follows;

$$J = \begin{pmatrix} 0 & 1 & 0 & 0 & 0 \\ 0 & 0 & 1 & 0 & 0 \\ 0 & 0 & 0 & 0 & 0 \\ 0 & 0 & 0 & 0 & 0 \\ 0 & 0 & 0 & 0 & 0 \end{pmatrix}.$$

5.2.8 **Example**

We will show that any nxn matrix A is similar to its transpose.

Let J be the Jordan form of A so that A is similar to J. Thus $P^{-1}AP = J$ for a suitable matrix P. Transposing this equation shows that A^t is similar to J^t since $(P^t)^{-1} = (P^{-1})^t$. It suffices then for us to show that J is similar to J^t.

First consider a Jordan block $J_\lambda = \begin{pmatrix} \lambda & 1 & 0 & 0 & & 0 \\ 0 & \lambda & 1 & 0 & & 0 \\ 0 & 0 & \lambda & 1 & & 0 \\ & & & & & \\ 0 & 0 & 0 & 0 & \lambda & 1 \\ 0 & 0 & 0 & 0 & 0 & \lambda \end{pmatrix}$.

$$\text{Let } P = \begin{pmatrix} 0 & 0 & 0 & 0 & & 0 & 1 \\ 0 & 0 & 0 & 0 & & 1 & 0 \\ 0 & 0 & 0 & & 1 & 0 & 0 \\ & & & & & & \\ 0 & 1 & 0 & 0 & & 0 & 0 \\ 1 & 0 & 0 & 0 & & 0 & 0 \end{pmatrix}$$

i.e. P has entries 1 on the anti-diagonal and zero elsewhere.

Then $P^2 = I$ so that $P^{-1} = P$ and an easy calculation will show that $P^{-1}J_{\lambda}P = J_{\lambda}^t$. Thus the result is true for a single Jordan block.

Finally for each Jordan block J_i we can construct a matrix P_i as above for which $P_i^{-1}J_iP_i = J_i^t$. Putting all of these matrices P_i together into a block matrix P we have that $P^{-1}JP = J^t$. This completes the proof.

5.2.9 **Comment**

As well as determining the Jordan form J of a matrix A it is sometimes necessary to determine a matrix P such that $P^{-1}AP = J$. The following example illustrates how to do this.

5.2.10 **Example**

$$A = \begin{pmatrix} -2 & -1 & -3 \\ 4 & 3 & 3 \\ -2 & 1 & -1 \end{pmatrix}$$

We met this matrix in Example (5.2.5) where we saw that its Jordan form is $J = \begin{pmatrix} -4 & 0 & 0 \\ 0 & 2 & 1 \\ 0 & 0 & 2 \end{pmatrix}$.

To find a matrix P such that $P^{-1}AP = J$ we denote the columns of P by p_1, p_2, and p_3 respectively. The equation $P^{-1}AP = J$ implies that $AP = PJ$ and equating columns from the two sides of this last equation yields the following three equations ;

$$Ap_1 = -4p_1$$

$$Ap_2 = 2p_2$$

$$Ap_3 = p_2 + 2p_3$$

Any eigenvector of A for the eigenvalue $\lambda = -4$ will be a solution of the first of these three equations. It is easily checked that $p_1 = \begin{pmatrix} 1 \\ -1 \\ 1 \end{pmatrix}$ is such an eigenvector.

Any eigenvector of A for the eigenvalue $\lambda = 2$ will be a solution of the second of these three equations. It is easily checked that $p_2 = \begin{pmatrix} 1 \\ -1 \\ -1 \end{pmatrix}$ is such an eigenvector.

Finally p_3 can be any solution of $Ap_3 = p_2 + 2p_3$ and we have already obtained $p_2 = \begin{pmatrix} 1 \\ -1 \\ -1 \end{pmatrix}$. Thus we have to solve $\begin{pmatrix} -2 & -1 & -3 \\ 4 & 3 & 3 \\ -2 & 1 & -1 \end{pmatrix}\begin{pmatrix} x \\ y \\ z \end{pmatrix} = \begin{pmatrix} 1 \\ -1 \\ -1 \end{pmatrix} + 2\begin{pmatrix} x \\ y \\ z \end{pmatrix}$.

The solution set of this system is $\left\{ \begin{pmatrix} 0 \\ 1 \\ 0 \end{pmatrix} + \alpha\begin{pmatrix} 1 \\ -1 \\ -1 \end{pmatrix} \; ; \alpha \in \mathbb{C} \right\}$ and so we can take $p_3 = \begin{pmatrix} 0 \\ -1 \\ 0 \end{pmatrix}$.

The matrix $P = \begin{pmatrix} 1 & 1 & 0 \\ -1 & -1 & -1 \\ 1 & -1 & 0 \end{pmatrix}$ will be a matrix for which $P^{-1}AP = J$. The reader should verify that this is the case. (Check that $AP = PJ$!)

5.2.11 **Remark**

In the special case when the matrix A is diagonalizable its Jordan form will consist entirely of $1{\times}1$ blocks and the columns of the matrix P will be a complete set of eigenvectors of A. In the more general situation, as in (5.2.10), the columns of P are either eigenvectors of A or "generalized eigenvectors", i.e. vectors such as p_3 in (5.2.10) which satisfies the equation $Ap_3 = p_2 + 2p_3$. A "generalized eigenvector" is a vector v which is the solution of an equation of the form $Av = w + \lambda v$ where w is a known vector and λ is an eigenvalue of the matrix A.

Problems 5A

1. If the $n \times n$ matrix A has trace zero and rank 1 show that A is nilpotent.

2. Determine the possible Jordan forms for an 8×8 matrix with characteristic polynomial $(\lambda - 6)^4(\lambda - 7)^4$ and minimal polynomial $(\lambda - 6)^2(\lambda - 7)^2$ if the eigenvalue $\lambda = 6$ has geometric multiplicity three.

3. Show that if the $n \times n$ matrix A satisfies $A^n = I$ then A is diagonalizable.

4. Let N be an $n \times n$ matrix which is nilpotent of index n. Use the binomial expansion of $(1 + x)^{1/2}$ to obtain a square root of $I + N$.
(i.e. obtain an $n \times n$ matrix B such that $B^2 = I + N$.)

5. Use the Jordan form together with problem 4 to show that any non-singular matrix in $M_n\mathbb{C}$ has a square root in $M_n\mathbb{C}$.

6. Find the Jordan form of $\begin{pmatrix} 2 & 0 & 0 & 0 & 0 & 0 \\ 1 & 2 & 0 & 0 & 0 & 0 \\ -1 & 0 & 2 & 0 & 0 & 0 \\ 0 & 1 & 0 & 2 & 0 & 0 \\ 1 & 1 & 1 & 1 & 2 & 0 \\ 0 & 0 & 0 & 0 & 1 & -1 \end{pmatrix}$.

7. Find the Jordan form of $\begin{pmatrix} 1 & 0 & 0 & 0 \\ 0 & 0 & 1 & 0 \\ 0 & 0 & 0 & 1 \\ 1 & 6 & -1 & -4 \end{pmatrix}$.

8. Let $A = \begin{pmatrix} 0 & 1 & 0 & 0 \\ 0 & 0 & 0 & 0 \\ 0 & 0 & 0 & 0 \\ 0 & 0 & 0 & 0 \end{pmatrix}$ and $B = \begin{pmatrix} 0 & 1 & 0 & 1 \\ 0 & 0 & 0 & 0 \\ 0 & 0 & 0 & 1 \\ 0 & 0 & 0 & 0 \end{pmatrix}$.

Show that A and B have the same characteristic polynomial, the same minimal polynomial, but not the same Jordan form.

9. If J is an $n{\times}n$ Jordan block with zero on the diagonal show that $J^n = 0$. If J_0 is any Jordan block with λ_0 on the diagonal and if p is the characteristic polynomial of J_0 deduce that $p(J_0) = 0$. Use this to provide another proof of the Cayley-Hamilton theorem.

10. Prove that the matrix $A \in M_n\mathbb{C}$ is diagonalizable if and only if the minimal polynomial of A is a product of distinct linear factors.

11. Let C be the companion matrix of the monic polynomial $p(x)$. (See problem 7 of Problems 4A and problem 6 of Problems 4C.) Deduce that $A \in M_n\mathbb{C}$ is similar to the companion matrix of its characteristic polynomial if and only if the characteristic and minimal polynomials of A are equal. (Strictly speaking we mean equal after multiplying by $(-1)^n$ to make the characteristic polynomial monic.)

5.3 Further remarks on the Jordan form

5.3.1 Determination of the Jordan form

The information given in (i), (ii), and (iii) of (5.2.2) is not in general sufficient to determine the Jordan form completely. However in lots of cases, especially for matrices which are not too large, it is in fact good enough. We will now give a method of completely determining the Jordan form of a matrix. It depends on a calculation of the ranks of the powers of the matrices $A - \lambda I$ for the various eigenvalues λ of the matrix.

Let $A \in M_n\mathbb{C}$ and let λ be an eigenvalue of A whose algebraic multiplicity is denoted by a_λ. Let k be the size of the largest block corresponding to λ and, for each $i = 1, 2,\ldots,k$, let N_i be the number of Jordan blocks of size i corresponding to λ in the Jordan form of A. Let $r_j = \text{rank } (A - \lambda I)^j$ for $j = 1, 2, \ldots$.

5.3.2 **Proposition**

(i) $a_\lambda = N_1 + 2N_2 + 3N_3 + \ldots\ldots + kN_k$.

(ii) $r_j = n - a_\lambda$ for all $j \geq k$ and $r_j > n - a_\lambda$ for all $j < k$.

(iii) $r_{k-1} = N_k + n \; a_\lambda$

$$r_{k-2} = 2N_k + N_{k-1} + n - a_\lambda$$

$$r_{k-3} = 3N_k + 2N_{k-1} + 3N_{k-2} + n - a_\lambda$$

$$\vdots$$

$$r_1 = (k-1)N_k + (k-2)N_{k-1} + \ldots\ldots + 2N_3 + N_2 + n - a_\lambda$$

Proof

Statement (i) follows at once from (i) in (5.2.2). We prove (ii) and (iii) as follows;

Since A is similar to its Jordan form J it follows that $A - \lambda I$ is similar to $J - \lambda I$ and that $(A - \lambda I)^j$ is similar to $(J - \lambda I)^j$ for each natural number j. Hence $r_j = \text{rank}\,(J - \lambda I)^j$ for each j as similar matrices have the same rank.

For a given eigenvalue λ of A an examination of $J - \lambda I$ shows that it consists of non-singular blocks for all eigenvalues different from λ and nilpotent blocks corresponding to λ.

The m×m matrix $B = \begin{pmatrix} 0 & 1 & 0 & 0 & 0 \\ 0 & 0 & 1 & 0 & 0 \\ 0 & 0 & 0 & 1 & 0 \\ 0 & 0 & 0 & 0 & 1 \\ 0 & 0 & 0 & 0 & 0 \end{pmatrix}$ has rank $m - 1$, the matrix B^2 has rank $m - 2$, B^3 has rank $m - 3$, etc., until we reach $B^m = 0$. Since $J - \lambda I$ has to be raised to the power k to kill off all of the nilpotent blocks, statement (ii) of (5.3.2) follows from this last observation about the ranks of the matrices B^j. Statement (iii) of (5.3.2) also follows from this.

5.3.3 **Remark**

We can use (5.3.2) in the following manner :

Given $A \in M_n\mathbb{C}$ we calculate a_λ for each eigenvalue λ of A. We then calculate $r_j = \text{rank } (A - \lambda I)^j$ for each j and each eigenvalue λ of A. The integer k will be the least value of j for which $r_j = n - a_\lambda$. Using (iii) we successively calculate $N_k, N_{k-1}, \ . \ , N_2$ and finally we use (i) to calculate N_1. See problem 2 of Problems 5B.

5.3.4 **Remark**

Sometimes a problem on matrices can be solved by transforming via a similarity to the Jordan form, solving in this case, and transforming the solution back via the inverse similarity.

5.3.5 **The real Jordan form**

For a matrix all of whose entries are real it is usual to view it as a complex matrix to obtain its Jordan form. However it is possible to have a variant of the Jordan form which has only real entries. It is known as the *real Jordan canonical form.* We refer the reader to [HJ, p152] for the details.

5.3.6 **The rational canonical form**

Any square matrix A can be shown to be similar to a direct sum of companion matrices of a certain set of polynomials called the invariant factors of A. We describe this briefly.

Consider the Jordan form J of the matrix A. Group together all of the Jordan blocks corresponding to a given eigenvalue. Do this for each eigenvalue. From each group choose a Jordan block of largest size and let B_1 be the direct sum of these chosen blocks. From what is left do the same again to obtain B_2. Keep on with this process until all of the blocks have been used up and we have obtained a set of matrices B_1, B_2, . . . , B_t for some integer t. Then the direct sum $\begin{pmatrix} B_1 & 0 & & 0 \\ 0 & B_2 & & 0 \\ & & & \\ 0 & 0 & & B_t \end{pmatrix}$ is similar to A as it is simply a permutation of the Jordan blocks.

For any Jordan block it is easy to see that the minimal and and characteristic polynomials are equal, (after multiplying by a factor -1 if necessary), and hence the same will be true of each matrix B_i. Now, by problem 11 of Problems 5A, each B_i is similar to the companion matrix of its minimal polynomial.

The sum of these companion matrices is called *the rational canonical form* of A. Let us write q_i for the minimal polynomial of the matrix B_i for each i = 1, 2, . . . ,t. The polynomials q_i are called *the invariant factors* of A. Observe that q_{i+1} is a divisor of q_i for each i = 1, 2, . . . , t-1 because of the manner of construction of the matrices B_i.

The rational canonical form can in fact be calculated without any reference to the Jordan form. It can be written down immediately once the invariant factors are known and these can be determined as follows ;

It can be shown that the matrix A - λI is always reducible to a diagonal matrix with entries $1,1, \ldots, 1, q_1, q_2, \ldots, q_t$, where each q_{i+1} divides q_i, via a sequence of elementary row and column operations of the kind met in chapter 1. Both row and column operations are permitted. These operations are required to be invertible and so we are allowed to add a polynomial multiple of one row or column to another but cannot divide by any non-constant polynomial. This amounts to the fact that there exist nxn matrices P and Q, whose entries are polynomials in λ and whose determinants are non-zero constants, such that PAQ is the diagonal matrix with entries $1,1, \ldots, 1, q_1, q_2, \ldots, q_t$.

The rational canonical form is so-called because the invariant factors, and hence the rational form itself, can be calculated in the above manner by a series of rational operations on the entries of the matrix. This procedure does not require the entries of the matrix to be in $\mathbb{C}$ but is valid for matrices with entries in $\mathbb{R}$ or indeed any field.

Problems 5B

1. Let A be an n×n matrix and let k be a natural number. Show that A^k tends to the zero matrix as k tends to infinity if and only if $|\lambda| < 1$ for each eigenvalue λ of A.

(*Hint* - first prove the result in the case when A is a Jordan block.)

2. Use the analysis of ranks, as in (5.3.2) and (5.3.3), to determine the Jordan form of the following matrix ;

$$\begin{pmatrix} 1 & 0 & 0 & 0 & 0 & 0 & 1 \\ 0 & 1 & 0 & 0 & 0 & 1 & 0 \\ 0 & 0 & 1 & 0 & 1 & 0 & 0 \\ 1 & 0 & 0 & 1 & 0 & 0 & 0 \\ 0 & 0 & 0 & 0 & 1 & 0 & 0 \\ 0 & 0 & 0 & 0 & 0 & 1 & 0 \\ 0 & 0 & 0 & 0 & 0 & 0 & 1 \end{pmatrix}$$

3. Show that the invariant factors of $\begin{pmatrix} 1 & 6 & 3 \\ 0 & 2 & 0 \\ 0 & 0 & 2 \end{pmatrix}$ are $(1-\lambda)(2-\lambda)$, and $2-\lambda$, and hence that the rational canonical form is $\begin{pmatrix} 2 & 0 & 0 \\ 0 & 0 & 1 \\ 0 & -2 & 3 \end{pmatrix}$.

Show that the only invariant factor of $\begin{pmatrix} 1 & 6 & 3 \\ 0 & 2 & 1 \\ 0 & 0 & 2 \end{pmatrix}$ is $(1-\lambda)(2-\lambda)^2$ and hence

that the rational canonical form is $\begin{pmatrix} 0 & 1 & 0 \\ 0 & 0 & 1 \\ 4 & -8 & 5 \end{pmatrix}$.

5.4 Systems of linear differential equations

5.4.1 The diagonal case

Let $x_1, x_2, \ldots, x_n$ each be real-valued functions of a single real variable t. (In applications t will usually be time.) We write $\dot{x}_i$ for the derivative of x_i with respect to t. If each $\dot{x}_i$ is expressible as a linear combination of $x_1, x_2, \ldots, x_n$ then we have a system of n linear differential equations of the form ;

$$\begin{aligned} \dot{x}_1 &= a_{11}x_1 + a_{12}x_2 + \ldots\ldots + a_{1n}x_n \\ \dot{x}_2 &= a_{21}x_1 + a_{22}x_2 + \ldots\ldots + a_{2n}x_n \\ \dot{x}_3 &= a_{31}x_1 + a_{32}x_2 + \ldots\ldots + a_{3n}x_n \\ &\;\vdots \\ \dot{x}_n &= a_{n1}x_1 + a_{n2}x_2 + \ldots\ldots + a_{nn}x_n \end{aligned}$$

where each $a_{ij} \in \mathbb{R}$.

If we write $x = \begin{pmatrix} x_1 \\ x_2 \\ \cdot \\ \cdot \\ x_n \end{pmatrix}$ and $\dot{x} = \begin{pmatrix} \dot{x}_1 \\ \dot{x}_2 \\ \cdot \\ \cdot \\ \dot{x}_n \end{pmatrix}$ then the above system of equations

can be written succinctly as a single matrix equation

$$\dot{x} = Ax$$

where $A = (a_{ij})$ is an $n \times n$ matrix with real entries.

If A is diagonal then $a_{ij} = 0$ for all $i \neq j$ and so we have n independent equations

$$\dot{x}_i = a_{ii}x_i, \qquad i = 1, 2, \ldots, n$$

which immediately yield the solutions

$$x_i = K_i e^{a_{ii}t}, \qquad i = 1, 2, \ldots, n$$

where K_i, $i = 1, 2, \ldots, n$ are constants which will be the values of x_i at $t = 0$. The constants K_i can thus be determined if the initial conditions are known.

5.4.2 **Proposition**

Let $A \in M_n\mathbb{R}$ be diagonalizable over $\mathbb{R}$ so that $P^{-1}AP = D$ where D is diagonal with real values $\lambda_1, \lambda_2, \ldots, \lambda_n$ on the diagonal and $P \in M_n\mathbb{R}$ is an invertible matrix. Then the system of linear differential equations $\dot{x} = Ax$ with the initial conditions given by the column vector $x(0) = x_0 \in \mathbb{R}^n$ has the solution $x = PD(t)P^{-1}x_0$ where $D(t)$ is the diagonal matrix with entries $e^{\lambda_1 t}, e^{\lambda_2 t}, \ldots, e^{\lambda_n t}$ on the diagonal.

Proof

Let the column vector $y = P^{-1}x$ so that $x = Py$. Then the equation $\dot{x} = Ax$ becomes $P\dot{y} = APy$ so that $\dot{y} = P^{-1}APy = Dy$.

The system has thus been changed to the simpler system with equations $\dot{y} = \lambda_i y_i$ for $i = 1, 2, \ldots, n$. This has solutions $y_i = K_i e^{\lambda_i t}$ for $i = 1, 2, \ldots, n$.

Putting $t = 0$ we have that $K_i = y_i(0)$ for each i and thus our solution $y = D(t)y(0)$. Substituting back yields $x = PD(t)P^{-1}x_0$ as required.

5.4.3 **Theorem**

Let $A \in M_n\mathbb{R}$ have Jordan form J and let $P^{-1}AP = J$ for some matrix P. then the system of linear differential equations $\dot{x} = Ax$ with initial conditions $x(0) = x_0$ has solution $x = PK(t)P^{-1}x_0$ where K(t) is the block diagonal matrix

$$\begin{pmatrix} K_1(t) & 0 & & 0 \\ 0 & K_2(t) & & \\ 0 & & & \\ & & & \\ 0 & & & K_r(t) \end{pmatrix}.$$

The blocks $K_i(t)$ correspond in a one-one fashion to the blocks in the Jordan form of A, each $K_i(t)$ being of the following form ;

$$K_i(t) = e^{\lambda_i t}\begin{pmatrix} 1 & t & t^2/2! & t^3/3! & & t^{m-1}/(m-1)! \\ 0 & 1 & t & t^2/2! & & t^{m-2}/(m-2)! \\ 0 & 0 & 1 & t & & t^{m-3}/(m-3)! \\ & & & & & \\ 0 & 0 & 0 & & 1 & t \\ 0 & 0 & 0 & & 0 & 1 \end{pmatrix}$$

corresponding to an mxm Jordan block with λ_i as eigenvalue.
($K_i(t)$ is upper triangular and is a *striped* matrix in the sense that it has the same entry $e^{\lambda_i t}$ at each point on the diagonal, the same entry $te^{\lambda_i t}$ at each point on the superdiagonal, the same entry $(t^2/2!)e^{\lambda_i t}$ at each point on the band above the superdiagonal etc.)

Proof

Putting $x = Py$ transforms the equation $\dot{x} = Ax$ into $\dot{y} = Jy$ in the same manner as in (5.4.2). The system $\dot{y} = Jy$ naturally breaks up into a set of smaller systems, one involving each Jordan block.

It suffices then to show that if $J = \begin{pmatrix} \lambda & 1 & 0 & 0 & & 0 \\ 0 & \lambda & 1 & 0 & & 0 \\ 0 & 0 & \lambda & 1 & & 0 \\ & & & & & \\ 0 & 0 & 0 & 0 & \lambda & 1 \\ 0 & 0 & 0 & 0 & 0 & \lambda \end{pmatrix}$, an m×m Jordan

block, then the solution of $\dot{y} = Jy$ is $y = K(t)y(0)$ where

$$K(t) = e^{\lambda t}\begin{pmatrix} 1 & t & t^2/2! & t^3/3! & & t^{m-1}/(m-1)! \\ 0 & 1 & t & t^2/2! & & t^{m-2}/(m-2)! \\ 0 & 0 & 1 & t & & t^{m-3}/(m-3)! \\ & & & & & \\ 0 & 0 & 0 & & 1 & t \\ 0 & 0 & 0 & & 0 & 1 \end{pmatrix}$$

The system $\dot{y} = Jy$ yields n equations

$$\dot{y}_1 = \lambda y_1 + y_2$$
$$\dot{y}_2 = \lambda y_2 + y_3$$
$$\vdots$$
$$\dot{y}_{n-1} = \lambda y_{n-1} + y_n$$
$$\dot{y}_n = \lambda y_n$$

Solving the last equation first gives $y_n = C_1e^{\lambda t}$ for some constant C_1. Substituting this solution into the next-to-the-last equation yields $\dot{y}_{n-1} = \lambda y_n + C_1e^{\lambda t}$. To solve this we multiply by the integrating factor $e^{-\lambda t}$ so that the equation may be rewritten as $d/dt(y_{n-1}e^{-\lambda t}) = C_1$. This gives a solution $y_{n-1} = (C_1t + C_2)e^{\lambda t}$ where C_2 is a constant.

Substitution of this value for y_{n-1} into the equation above gives $\dot{y}_{n-2} = \lambda y_{n-2} + (C_1t + C_2)e^{\lambda t}$. Using the same integrating factor $e^{-\lambda t}$

this equation becomes $d/dt(y_{n-2}e^{-\lambda t}) = C_1t + C_2$. This gives a solution $y_{n-2} = (C_1(t^2/2) + C_2t + C_3)e^{\lambda t}$ where C_3 is a constant.

Proceeding in this fashion we eventually obtain the result that $y = K(t)y(0)$ where $y(0)$ is the column vector $\begin{pmatrix} C_n \\ C_{n-1} \\ \vdots \\ C_1 \end{pmatrix}$. This completes the proof.

(Notice that in the proof of (5.4.3) and (5.4.2) the substitution $x = Py$ has the effect of disentangling the variables and giving a new system which is easy to solve.)

5.4.4 **Example**

Solve the system $\dot{x} = Ax$ where $A = \begin{pmatrix} -2 & -1 & -3 \\ 4 & 3 & 3 \\ -2 & 1 & -1 \end{pmatrix}$ with the initial condition $x(0) = \begin{pmatrix} 1 \\ 0 \\ 1 \end{pmatrix}$.

We saw in (5.2.5) that A has Jordan form $J = \begin{pmatrix} -4 & 0 & 0 \\ 0 & 2 & 1 \\ 0 & 0 & 2 \end{pmatrix}$, and in (5.2.10) we saw that $P^{-1}AP = J$ where $P = \begin{pmatrix} 1 & 1 & 0 \\ -1 & -1 & -1 \\ 1 & -1 & 0 \end{pmatrix}$.

It is straightforward to calculate that $P^{-1} = \begin{pmatrix} 1/2 & 0 & 1/2 \\ 1/2 & 0 & -1/2 \\ -1 & -1 & 0 \end{pmatrix}$.

From (5.4.3) the solution to the system will be $x = PK(t)P^{-1}x(0)$ where

$$K(t) = \begin{pmatrix} e^{-4t} & 0 & 0 \\ 0 & e^{2t} & te^{2t} \\ 0 & 0 & e^{2t} \end{pmatrix}.$$

$$\text{Thus } x = \begin{pmatrix} 1 & 1 & 0 \\ -1 & 1 & -1 \\ 1 & -1 & 0 \end{pmatrix} \begin{pmatrix} e^{-4t} & 0 & 0 \\ 0 & e^{2t} & te^{2t} \\ 0 & 0 & e^{2t} \end{pmatrix} \begin{pmatrix} 1/2 & 0 & 1/2 \\ 1/2 & 0 & -1/2 \\ -1 & -1 & 0 \end{pmatrix} \begin{pmatrix} 1 \\ 0 \\ 1 \end{pmatrix}$$

$$= \begin{pmatrix} e^{-4t} - te^{2t} \\ -e^{-4t} - te^{2t} + e^{t} \\ e^{-4t} + te^{2t} \end{pmatrix}$$

5.4.5 **Remark**

The method of (5.4.3) may well yield a complex solution since some of the eigenvalues of the matrix A could be complex and not real. In that case we must take the real part of this complex solution to obtain the solution of the original problem. (All coefficients and functions in the original problem are real !) The technique we have described is also valuable for solving systems of the form $\dot{x} = Ax + g$ where g is a column vector of functions, and also for second order systems such as $\ddot{x} = Ax$. See problems 5C for details of this.

5.4.6 **The solution via matrix exponentials**

The solution of the system $\dot{x} = Ax$ as given in (5.4.3) may equivalently be expressed in terms of the matrix exponential which we have encountered already in (3.5.2).

If J is the Jordan form of the matrix A then $P^{-1}AP = J$ for a suitable matrix P so that $A = PJP^{-1}$. Hence $A^k = PJ^kP^{-1}$ for each natural number k and then $\exp A = P (\exp J) P^{-1}$.

Now suppose first that J is an m×m Jordan block ;

$$J = \begin{pmatrix} \lambda & 1 & 0 & 0 & & 0 \\ 0 & \lambda & 1 & 0 & & 0 \\ 0 & 0 & \lambda & 1 & & 0 \\ & & & & & \\ 0 & 0 & 0 & 0 & \lambda & 1 \\ 0 & 0 & 0 & 0 & 0 & \lambda \end{pmatrix}$$

Calculating the powers J^2, J^3, etc. and using the definition of exp J as a power series leads to the following expression ;

$$\exp J = e^{\lambda}\begin{pmatrix} 1 & 1 & 1/2! & 1/3! & & 1/(m-1)! \\ 0 & 1 & 1 & 1/2! & & 1/(m-2)! \\ 0 & 0 & 1 & 1 & & 1/(m-3)! \\ & & & & & \\ & & & & & \\ 0 & 0 & 0 & & 1 & 1 \\ 0 & 0 & 0 & & 0 & 1 \end{pmatrix}.$$

(The reader may find it helpful to do the above calculation first for 2x2 blocks, and then for 3x3 blocks.)

Notice that the above matrix exp J is precisely the matrix obtained by putting t = 1 into the matrix K(t) of (5.4.3).

Multiplying each entry of the matrix J by t and doing a similar calculation we find that exp (tJ) = K(t).

Now if the Jordan form of the matrix A is made up of blocks J_1, J_2, . . . ,J_r then exp J is made up of blocks exp J_1, exp J_2, ... , exp J_r. Thus exp (tA) = P (exp (tJ))P^{-1} = P K(t) P^{-1} so that the solution of the system $\dot{x} = Ax$ can be rewritten in the form x = (exp tA) x_0. This may be regarded as a generalization of the single variable differential equation $\dot{x} = ax$, with $a \in \mathbb{R}$, that equation having solution $x = x_0 e^{at}$.

5.4.7 Stability of solutions

The solution x(t) of a system of linear differential equations $\dot{x} = Ax$ is said to be *unstable* if x(t) tends to infinity as t tends to infinity. The solution is said to be *stable* if x(t) tends to zero as t tends to infinity. (If x(t) is bounded as t tends to infinity the solution is sometimes said to be *neutrally stable*.)

The following gives conditions for stability ;

5.4.8 Proposition

(i) The solution x(t) of the system $\dot{x} = Ax$ tends to zero as t tends to infinity if and only if, for each eigenvalue λ of A, the real part of λ is negative. (i.e. the eigenvalues of A all lie in the left-hand half of the complex plane.)

(ii) The solution x(t) is bounded, i.e. there exists a constant K such that $|x(t)| \leq K$ for all t, if and only if, for each eigenvalue λ of A, the real part of λ is non-positive and for any λ whose real part is zero the algebraic and geometric multiplicities of λ coincide.

Proof

(i) The solution can be written $x(t) = (\exp tA)x_0$. From (5.4.6) it is clear that exp tA tends to zero as t tends to infinity if and only if exp tJ tends to zero, J being the Jordan form of A. The entries of exp tJ are all of the form $e^{\lambda t}p(t)$ where p is a polynomial in t. Letting $\lambda = r + is$ we have $|e^{\lambda t}| = e^{rt}$ since $|e^{ist}| = 1$. It is well-known from real analysis that, for any polynomial p(t), $e^{rt}p(t)$ tends to zero as t tends to infinity if and only if $r < 0$. This proves (i).

(ii) Any non-constant polynomial p(t) is unbounded as t tends to infinity. If the real part of the eigenvalue λ of A is zero and if the algebraic and geometric multiplicities of λ coincide then all of the Jordan blocks in J corresponding to λ must be 1x1 blocks. Hence the blocks in exp tJ corresponding to λ will be 1x1 blocks of the form (e^{ist}) and these have modulus one. All possibilities for λ other than those mentioned already will lead to exp tJ being unstable.

5.5 **Markov processes**

This topic was mentioned briefly in (1.7(g)) and (4.7(b)). We now go into detail about it.

5.5.1 **Definition**

An n×n matrix A with real entries a_{ij} is called a *stochastic matrix* provided that $a_{ij} \geq 0$ for all i, j and $\sum_{i=1}^{n} a_{ij} = 1$ for each j = 1, 2, . . n, (i.e. each entry of A is non-negative and, for each column of A, the sum of the entries in that column is one.)

If also each entry of A is positive then A is called a *positive stochastic matrix.*

5.5.2 **Remark**

Stochastic matrices occur in probability theory, the entries of the matrix usually being the probabilities of different events. A stochastic matrix is called *doubly stochastic* if also the sum of the entries in each row is one.

5.5.3 **Definition**

Let $x = (x_i)$ be a column vector in $\mathbb{R}^n$ with $\sum_{i=1}^{n} x_i = 1$ and $x_i \geq 0$ for each i. Let A be a stochastic n×n matrix.

The sequence of vectors $\{ A^k x \}$, k = 1, 2, ... , is called a *finite Markov process,* (or *Markov chain*).

The matrix A is called the *transition matrix* of the process.

The vector x is called the *initial state vector* and the vector $A^k x$ is called the *state vector after k steps* of the process.

5.5.4 **Example**

Three companies X, Y, and Z simultaneously introduce a new home computer on the market. They start with equal shares of the market. Each month company X loses 5% of its customers to company Y, 10% of its customers to company Z, but retains the other 85% . Company Y retains 75% of its customers each month but loses 15% to company X and 10% to company Z. Company Z retains 90% of its customers but loses 5% to each of the other two companies. We will formulate this as a Markov process.

The first column of the transition matrix A has as its entries the probabilities that a client of company X will be a client of companies X, Y, and Z respectively one month later. The second and third columns of A have similar entries for the clients of Y and Z.

$$A = \begin{pmatrix} \frac{85}{100} & \frac{15}{100} & \frac{5}{100} \\ \frac{5}{100} & \frac{75}{100} & \frac{5}{100} \\ \frac{10}{100} & \frac{10}{100} & \frac{90}{100} \end{pmatrix}$$

The initial state $x = \begin{pmatrix} 1/3 \\ 1/3 \\ 1/3 \end{pmatrix}$ since the three companies initially have equal shares of the market.

5.5.5 **Exercise**

In the above example what share of the market will be held by company Z after three months ? (Calculate the third component of the vector A^3x.)

5.5.6 **Geometric lemma**

Let $z_1, z_2, \ldots, z_n$, be a set of complex numbers with $|z_j| = 1$ for each j. Let $\alpha_1, \alpha_2, \ldots, \alpha_n$ be real numbers for which $0 < \alpha_j < 1$ for each j and $\sum_{j=1}^{n} \alpha_j = 1$. Suppose further that $|\sum_{j=1}^{n} \alpha_j z_j| = 1$. Then all of the z_j must be equal.

Proof

We will assume that not all z_j are equal and deduce a contradiction.

Firstly note that there is no loss of generality in assuming that the z_j are all different. (For example if $z_1 = z_2$ then we could omit z_2 and replace the coefficient α_1 by a new coefficient $\alpha_1 + \alpha_2$.)

The z_j, $j = 1, 2, \ldots, n$, are the vertices of a polygon inscribed in the unit circle. Now $\sum_{j=1}^{n} \alpha_j z_j$ lies somewhere on this polygon because $\sum_{j=1}^{n} \alpha_j = 1$. Also, because $0 < \alpha_j < 1$ for each j, the point $\sum_{j=1}^{n} \alpha_j z_j$ cannot actually be one of the vertices. However, since $|\sum_{j=1}^{n} \alpha_j z_j| = 1$, the point $\sum_{j=1}^{n} \alpha_j z_j$ must lie on the unit circle. This is a contradiction and so the lemma is proven.

5.5.7 **Theorem**

Let A be a positive stochastic matrix. Then the real number one occurs as an eigenvalue of A with algebraic multiplicity one. All other eigenvalues λ of A satisfy $|\lambda| < 1$.

Proof

Using (4.4.2) with the maximum absolute sum norm it follows that $|\lambda| \leq 1$ for each eigenvalue λ of A. Now suppose that λ is an eigenvalue of A with $|\lambda| = 1$. We will show that $\lambda = 1$.

Let z be an eigenvector for the matrix A^t corresponding to the eigenvalue λ. Recall from (4.2.4) that A and A^t have the same eigenvalues ! We write $z = (z_j)$, a column vector, and choose z so that $\max_j |z_j| = 1$. This is always possible since any non-zero scalar multiple of an eigenvector remains an eigenvector.

Choose j_0 such that $|z_{j_0}| = 1$ and $|z_j| \leq 1$ for all j. The equation $A^t z = \lambda z$ yields a set of n equations and the j_0-th equation of this set is as follows ;

$$a_{1j_0} z_1 + a_{2j_0} z_2 + \ldots\ldots + a_{nj_0} z_n = \lambda z_{j_0}$$

Hence $\left| \sum_{k=1}^{n} a_{kj_0} z_k \right| = 1$ since $|\lambda| = 1$ and $|z_{j_0}| = 1$. Now, from the triangle inequality, we see that $1 \leq \sum_{k=1}^{n} |a_{kj_0}| \, |z_k|$.

Also $1 = \left| \sum_{k=1}^{n} a_{kj_0} \right| \, |z_{j_0}|$ from the fact that A is stochastic and from the choice of j_0. Thus $\left| \sum_{k=1}^{n} a_{kj_0} \right| \, |z_{j_0}| \leq \sum_{k=1}^{n} |a_{kj_0}| \, |z_k|$ which yields $0 \leq \sum_{k=1}^{n} |a_{kj_0}| \, (|z_k| - |z_{j_0}|)$.

Since A is positive and $|z_k| \leq |z_{j_0}|$ for all k this last inequality is impossible unless $|z_k| = |z_{j_0}|$ for all k.

Applying (5.5.6) to the set of complex numbers z_k and the real numbers a_{kj_0} we find that all of the z_k must be equal.

Hence, for some $\beta \in \mathbb{C}$ with $|\beta| = 1$, we have $z_k = \beta$ for all k and so $z = \beta e$ where e is the column vector with every component equal to one. Now the equation $A^t z = \lambda z$ implies that $A^t \beta e = \lambda \beta e$. However, by linearity, $A^t \beta e = \beta A^t e = \beta e$. (Note that $A^t e = 1$ because A is stochastic.) This implies that $\lambda = 1$ since $\beta \neq 0$ and $e \neq 0$.

Finally we must prove that the eigenvalue $\lambda = 1$ has algebraic multiplicity one. From the above it follows that the eigenvalue $\lambda = 1$ for the matrix A^t has geometric multiplicity one. We saw in (5.2.7) that the matrices A and A^t have the same Jordan form and so their eigenvalues must have the same geometric multiplicities. The Jordan form J of A will thus have only one block corresponding to the eigenvalue one and it suffices for us to show that this block is a 1x1 block.

Now $P^{-1}AP = J$ for some matrix P and therefore $J^m = PA^mP^{-1}$ for all natural numbers m. Using the maximum absolute column sum norm, which is an operator norm and thus satisfies (*) of (3.3.6), it is clear that $\|A^m\| < 1$ for all m. Using properties of operator norms we have $\|J^m\| \leq \|P\| \ \|A^m\| \ \|P^{-1}\| \leq \|P\| \ \|P^{-1}\|$. This shows that $\{\|J^m\|\}$ is a bounded set. However if J contains a block of size greater than one corresponding to $\lambda = 1$ then $\{\|J^m\|\}$ cannot be bounded. (It is an easy exercise to show that $\|J^m\| > m$ if J contains such a block.) This completes the proof.

5.5.8 **Corollary**

Let A be a positive stochastic matrix. Then there exists a vector $v \in \mathbb{R}^n$ with the property that, for all vectors $v_0 \in \mathbb{R}^n$, the sequence $\{A^k v_0\}$ tends to some scalar multiple of v as k tends to infinity. This vector v is unique up to a scalar multiple and does not depend on v_0. The vector v can be chosen to have all of its components positive.

Proof

Let v be an eigenvector for A corresponding to the eigenvalue $\lambda = 1$. The Jordan form J of A can be written with a 1x1 block for $\lambda = 1$ in the top left-hand corner and $|\lambda| < 1$ for all the other blocks.

Following the method in (5.2.10) a matrix P such that $P^{-1}AP = J$ can be chosen to have the eigenvector v as its first column. Then $A^k = PJ^kP^{-1}$ so that $A^k v_0 = PJ^kP^{-1}v_0$. As k tends to infinity J^k tends to a matrix with 1 in the top left-hand corner and zero elsewhere. (This is because all other Jordan blocks in J have $|\lambda| < 1$.) Thus PJ^k tends to a matrix with v as its first column and zero elsewhere. Writing $P^{-1}v_0$ as a column vector (y_i) we see that $A^k v_0 = PJ^kP^{-1}v_0$ tends to $y_1 v$ as k tends to infinity. If e is the vector in (5.5.7) then all of the components of $A^k e$ are positive. Taking $v_0 = e$ thus proves the final statement of the corollary and completes the proof.

5.5.9 **Comment**

The above corollary shows that the Markov process tends to a unique *steady state* described by the vector v, and this steady state is the same for all choices of the initial state v_0. We often say that the steady state vector v describes what happens "in the long run".

Note that the steady state vector is the solution of the equation Av = v, i.e. an eigenvector corresponding to $\lambda = 1$ for the matrix A. The vector e in the proof of (5.5.8) is an eigenvector corresponding to $\lambda = 1$ for the matrix A^t but not in general for A.

5.5.10 **Example**

For the example of (5.5.4) we will determine how much of the market each of the companies X, Y, and Z has in the long run.

In (5.5.4) the transition matrix A was as follows ;

$$A = \begin{pmatrix} \frac{85}{100} & \frac{15}{100} & \frac{5}{100} \\ \frac{5}{100} & \frac{75}{100} & \frac{5}{100} \\ \frac{10}{100} & \frac{10}{100} & \frac{90}{100} \end{pmatrix}$$

Solving $A \begin{pmatrix} x \\ y \\ z \end{pmatrix} = \begin{pmatrix} x \\ y \\ z \end{pmatrix}$ we obtain the three equations

$$85x + 15y + 5z = 100x$$

$$5x + 75y + 5z = 100y$$

$$10x + 10y + 90z = 100z$$

The solution set of this set of linear equations is easily checked to be $\{\alpha \begin{pmatrix} 2 \\ 1 \\ 3 \end{pmatrix} : \alpha \in \mathbb{R} \}$. We usually choose the steady state vector to be normalized, (i.e. the sum of the components equals one), and so in this case we take

$v = \begin{pmatrix} 1/3 \\ 1/6 \\ 1/2 \end{pmatrix}$. In the long run the companies X, Y, and Z will have respectively 1/3, 1/6, and 1/2 of the market.

5.5.11 **Remark**

Theorem (5.5.7) and Corollary (5.5.8) require the stochastic matrix A to be positive. If A is stochastic but has some entries equal to zero then the conclusions of (5.5.7) and (5.5.8) need not hold. For example consider

$A = \begin{pmatrix} 1 & 0 & 0 \\ 0 & 1/2 & 1/4 \\ 0 & 1/2 & 3/4 \end{pmatrix}$ which has eigenvalue one occurring with algebraic multiplicity two and geometric multiplicity two. What happens in the long run will depend on the initial state.

5.5.12 **Definition**

The stochastic matrix A is called a *regular* stochastic matrix provided that there is some natural number k such that all of the entries of A^k are positive.

5.5.13 **Proposition**

Let A be a regular stochastic matrix. Then all of the conclusions of Theorem (5.5.7) and Corollary (5.5.8) are valid for the matrix A.

Proof

Let $P^{-1}AP = J$ where J is a Jordan form for A. Then J^k has the same Jordan form as A^k since J^k and A^k are similar. Since A^k is positive we can apply (5.5.7) to A^k. Thus the Jordan form of J^k contains a 1x1 block with $\lambda = 1$ and all other blocks of J^k have $|\lambda| < 1$. The same statement must hold for the blocks of J because the diagonal entries of J^k are the k-th powers of the diagonal entries of J. Hence all of the conclusions of (5.5.7) and (5.5.8) are valid for the regular stochastic matrix A.

In particular a Markov process with a regular stochastic transition matrix will have a unique steady state.

5.5.14 **Remark**

Theorem (5.5.7) may be regarded as a special case of the theorem of Perron. This theorem says that if $A \in M_n\mathbb{R}$ has all positive entries and $r = \max \{ |\lambda| : \lambda \in \sigma(A) \}$ then r occurs as an eigenvalue of A with algebraic multiplicity one and $|\lambda| < r$ for all other $\lambda \in \sigma(A)$. Furthermore, as in (5.5.8), there is an eigenvector for r with all of its components positive.

More generally there is the Perron-Frobenius theorem which says that if $A \in M_n\mathbb{R}$ is irreducible, (see below for the definition), and has all non-negative entries then the same conclusions hold with the modification that $|\lambda| \leq r$ for all other $\lambda \in \sigma(A)$.

(A is said to be *reducible* if there exists a permutation of the rows of A such that, when combined with the same permutation of the columns, A is reduced to the form $\begin{pmatrix} B & 0 \\ C & D \end{pmatrix}$ for some matrices B, C, and D, i.e. if $P^tAP = \begin{pmatrix} B & 0 \\ C & D \end{pmatrix}$ for some permutation matrix P. Otherwise A is said to be *irreducible*.)

See [Se] for further information.

5.6 **The Sylvester matrix equation**

Let A, B, and C be fixed n×n matrices. Suppose that we want to find an n×n matrix X such that $AX + XB = C$. An equation of this type is known as a *Sylvester matrix equation.* Such equations are encountered in particular by engineers in control theory. From one viewpoint solving a Sylvester matrix equation amounts simply to solving the system of n^2 linear equations in n^2 unknowns, the n^2 entries of the matrix X which we are seeking. This can be solved by Gaussian elimination as we have seen in chapter 1. However it is valuable to reduce the problem to that of solving n smaller systems of linear equations, each smaller system being a set of n linear equations in n unknowns. The technique is as follows.

Suppose first that the matrix B is diagonalizable, i.e. there is a diagonal matrix D and a non-singular matrix Q such that $Q^{-1}BQ = D$. Multiplying the equation $AX + XB = C$ on the right by Q and putting $E = XQ$ reduces the problem to finding an $n \times n$ matrix E such that $AE + ED = CQ$, i.e. we have reduced the problem to solving the Sylvester matrix equation in the case when B is diagonal.

Let $e_1, e_2, \ldots, e_n$ be the columns of E. Each e_i is a column vector of n unknowns. By problem 3 of Problems 1A the matrix AE has as its columns the vectors $Ae_1, Ae_2, \ldots, Ae_n$. The matrix ED has as its columns $\lambda_1 e_1, \lambda_2 e_2, \ldots, \lambda_n e_n$ where the λ_i are the diagonal entries of D. If we write $t_1, t_2, \ldots, t_n$ for the columns of the matrix CQ then the equation $AE + ED = CQ$ yields, on equating columns on either side of the equation, the set of n equations ;

$$\begin{aligned} Ae_1 + \lambda_1 e_1 &= t_1 \\ Ae_2 + \lambda_2 e_2 &= t_2 \\ &\cdot \\ &\cdot \\ Ae_n + \lambda_n e_n &= t_n \end{aligned}$$

The first equation is a set of n linear equations in n unknowns, the entries of e_1, and solving this system gives e_1. Similarly the other equations give $e_2, e_3, \ldots, e_n$. Having obtained E we can then calculate X from $X = EQ^{-1}$.

In general we can only reduce B to a Jordan form J via a transformation $Q^{-1}BQ = J$. As earlier the Sylvester matrix equation reduces to the equation

AE + EJ = CQ where E = XQ. If J is a single Jordan block $\begin{pmatrix} \lambda & 1 & 0 & 0 & & 0 \\ 0 & \lambda & 1 & 0 & & 0 \\ 0 & 0 & \lambda & 1 & & 0 \\ & & & & & \\ 0 & 0 & 0 & 0 & \lambda & 1 \\ 0 & 0 & 0 & 0 & 0 & \lambda \end{pmatrix}$ then

this matrix equation yields n equations

$$Ae_1 + \lambda e_1 = t_1$$

$$Ae_2 + e_1 + \lambda e_2 = t_2$$

$$\cdot$$

$$\cdot$$

$$Ae_n + e_{n-1} + \lambda e_n = t_n$$

The first equation is a system of n linear equations for the n unknown entries of e_1. Solve for e_1, substitute into the second equation and solve for e_2, etc. In this way E is obtained and finally $X = EQ^{-1}$. In general J will be a sum of Jordan blocks and each block is dealt with in the above manner.

It is often desirable to know when the Sylvester matrix equation has a unique solution. The following proposition gives information on this. First we recall some notation from chapter 4.

The spectrum $\sigma(A) = \{\lambda \in \mathbb{C} : \lambda \text{ is an eigenvalue of } A\}$. Also we write $-\sigma(B) = \{-\lambda : \lambda \in \sigma(B)\}$.

5.6.1 **Proposition**

The Sylvester matrix equation AX + XB = C has a unique solution if and only if $\sigma(A) \cap -\sigma(B)$ is the empty set.

(This condition means that if λ is an eigenvalue of A then $-\lambda$ cannot be an eigenvalue of B.)

Proof

Observe first that AX + XB = C has a unique solution if and only if

$AX + XB = 0$ has a unique solution. (Note that $X = 0$ is always a solution of $AX + XB = 0$. If $AX + XB = C$ has a unique solution then the coefficient matrix of this system of n^2 linear equations must be non-singular. Consequently $AX + XB = 0$, the associated homogeneous system with the same coefficient matrix, must have the unique solution $X = 0$. If, on the other hand, the system $AX + XB = C$ fails to have a unique solution, either by having no solution or by having many solutions, then the above coefficient matrix is necessarily singular and hence $AX + XB = 0$ will have a non-zero solution.)

Solving $AX + XB = 0$ by performing a reduction of the kind described before this proposition we have a set of equations as above with $t_i = 0$ for all i. Examining these equations it can be seen that a non-zero solution for them will exist if and only if there is some eigenvalue λ of A for which $-\lambda$ is an eigenvalue of B. This completes the proof.

5.6.2 **The Liapunov stability criterion**

An important special case of the Sylvester equation is a matrix equation of the form $\bar{A}^t X + XA = -I$. It is known as the *Liapunov matrix equation*. It leads to a criterion for the stability of the solution of a system $\dot{x} = Ax$ of linear differential equations.

We have seen in (5.4.8) that the system $\dot{x} = Ax$ of linear differential equations has a solution which is stable if and only if all of the eigenvalues of the matrix A lie in the left-hand half of the complex plane.

We say that the matrix A is *stable* if its eigenvalues all lie in the left-hand half plane.

Arising from the classical theory of Liapunov on stability of motion there is a criterion for the stability of A which depends on solutions of the Liapunov matrix equation.

This criterion has also been used in mathematical economics as a measure of the stability of an economic system.

An $n \times n$ hermitian matrix H is said to be *positive-definite* provided that $\bar{v}^t H v$ is a positive real number for all non-zero vectors $v \in \mathbb{C}^n$. The notion of positive-definiteness will be done in detail in chapter 6. For the moment we need only the definition and the fact that if H is positive-definite then $H = P\bar{P}^t$ for some invertible matrix P. (See problem 6 of Problems 6C for this.)

5.6.3 **Proposition**

The $n \times n$ matrix A is stable if and only if there exists a positive-definite matrix which is a solution of the Liapunov matrix equation

$$\bar{A}^t X + XA = -I.$$

Proof

Assume that there exists a positive- definite matrix H satisfying $\bar{A}^t H + HA = -I$. Then we can write $H = P\bar{P}^t$ for some invertible matrix P. Putting $H = P\bar{P}^t$ and multiplying the Liapunov equation on the left by P^{-1} and on the right by $(\bar{P}^t)^{-1}$ yields that $\bar{Y}^t + Y = -Q\bar{Q}^t$ where $Y = \bar{P}^t A(\bar{P}^t)^{-1}$ and $Q = P^{-1}$.

Let λ be an eigenvalue of A so that λ is also an eigenvalue of Y since A and Y are similar. Then $Yv = \lambda v$ for some non-zero vector v. Taking the conjugate transpose gives $\bar{v}^t \bar{Y}^t = \bar{\lambda}\bar{v}^t$ and from this we see that $\bar{v}^t(\bar{Y}^t + Y)v = (\bar{\lambda} + \lambda)\bar{v}^t v$.

But $\bar{Y}^t + Y = -Q\bar{Q}^t$ so that this last equation may be rewritten as $-\|\bar{Q}^t v\|^2 = (\bar{\lambda} + \lambda)\|v\|^2$, $\| \ \|$ being the Euclidean norm. Thus $(\bar{\lambda} + \lambda) < 0$ and, since $\bar{\lambda} + \lambda = 2(\mathrm{Re}\ \lambda)$, all the eigenvalues of A must lie in the left-hand half-plane. This shows that A is stable.

Conversely assume that A is stable. Then, by (5.4.8), any solution x(t) of $\dot{x} = Ax$ tends to zero as t tends to infinity. Let B be any n×n hermitian matrix and let $g_B(t) = \overline{x(t)}^t Bx(t)$, which is a real valued function of the real variable t. (Remember that x(t) is a column vector of length n.) Then $g_B(t)$ tends to zero as t tends to infinity.

The Liapunov matrix equation $\bar{A}^t X + XA = -I$ will have a unique solution when A is stable by (5.6.1). (Note that the eigenvalues of $\bar{A}^t$ are the complex conjugates of those of A.) The uniqueness of the solution implies that $X = \bar{X}^t$ because transposing the Liapunov equation shows that $\bar{X}^t$ is also a solution. We write H for this unique solution and we will show by a contradiction argument that H is necessarily positive-definite.

Suppose there exists a vector $v_0 \in \mathbb{C}^n$ such that $\bar{v}_0^t H v_0 \leq 0$. For the system $\dot{x} = Ax$ take as initial condition $x(0) = v_0$. Taking B = H we see that $g_H(0) = \bar{v}_0^t H v_0 \leq 0$. Using the product rule for differentiation of $g_H(t)$ we find that

$$\frac{d}{dt}(g_H(t)) = \overline{(\dot{x}(t))}^t \, H \, x(t) + \overline{x(t)}^t \, H \, \dot{x}(t)$$

$$= \overline{x(t)}^t \, (\bar{A}^t H + HA) \, x(t)$$

$$= -\overline{x(t)}^t \, x(t).$$

(using $\dot{x} = Ax$ and $\bar{A}^t H + HA = -I$.)

This shows that the function $g_H(t)$ is strictly decreasing because its derivative is always negative. This yields the desired contradiction since $g_H(0) \leq 0$ and $g_H(t)$ tends to zero as t tends to infinity.

Problems 5C

1. Solve the system of linear differential equations

$$\dot{x}_1 = x_1 + 4x_2$$

$$\dot{x}_2 = x_1 + x_2$$

with initial conditions $x_1(0) = 4$, $x_2(0) = 8$.

2. Solve the system of linear differential equations

$$\dot{x}_1 = 2x_1 + x_2 + x_3$$
$$\dot{x}_2 = 2x_2 + 2x_3$$
$$\dot{x}_3 = 3x_3$$

with initial conditions $x_1(0) = 2$, $x_2(0) = 0$, $x_3(0) = 1$.

3. Solve the system of second-order linear differential equations

$$\ddot{x} = -9x + 24y$$
$$\ddot{y} = -4x + 11y$$

4. Solve the system of differential equations

$$\dot{x}_1 = 2x_1 - x_2 + e^t$$
$$\dot{x}_2 = -2x_1 + 2x_2 + 1$$

5. If a mother has red hair the probability that her child has red hair is 0.6, that her child has blonde hair is 0.3, and brown hair 0.1. The children of blonde-haired mothers have probabilities of 0.1, 0.7, and 0.2 for red, brown, and blonde hair respectively. The children of brown-haired mothers split 0.3, 0.4, and 0.3 for red , brown, and blonde respectively.

Assuming that the population consists only of people with these three hair colours set up a transition matrix for a Markov process and find the long-term percentage of people with red hair.

6. A student studies the 3 subjects mathematics, physics and chemistry. He never studies the same subject on successive days. If he studies mathematics one day then next day he studies physics. If he studies either physics or chemistry one day then next day he is twice as likely to study mathematics as the other subject.

Write down the transition matrix for this process and show that it is regular.

If he studies mathematics on the first day what is the probability that he studies chemistry on the fourth day ?

In the long run how often does he study each subject ?

7. The continent of Atlantis is divided into three regions, Upper, denoted U Middle, denoted M, and Lower, denoted L. The yearly population movement between the regions obeys the following transition matrix ;

$$\begin{array}{c} \\ U \\ M \\ L \end{array}\begin{array}{c} \begin{array}{ccc} U & M & L \end{array} \\ \begin{pmatrix} 0.4 & 0.1 & 0.1 \\ 0.5 & 0.7 & 0.5 \\ 0.1 & 0.2 & 0.4 \end{pmatrix} \end{array}$$

Given an initial population distribution of 20% in U, 70% in M, and 10% in L determine the percentage of the population that will be in the region M after two years.

What percentage of the population will be in the region M in the long run ?

8. A Markov process has the following transition matrix ;

$$\begin{pmatrix} 0 & 0.2 & 0 & 0 & 0 & 0 \\ 1 & 0 & 0.4 & 0 & 0 & 0 \\ 0 & 0.8 & 0 & 0.6 & 0 & 0 \\ 0 & 0 & 0.6 & 0 & 0.8 & 0 \\ 0 & 0 & 0 & 0.4 & 0 & 1 \\ 0 & 0 & 0 & 0 & 0.2 & 0 \end{pmatrix}$$

Show that A is not regular and that in the long run the system oscillates between two states.

9. Let A be an $n \times n$ stochastic matrix. If fewer than $n/2$ entries in each row of A are zero and fewer than $n/2$ entries in each column of A are zero show that A is regular.

10. Solve the Sylvester matrix equation in the following two cases

(i) $A = \begin{pmatrix} -1 & 1 \\ 0 & -1 \end{pmatrix}$, $B = \begin{pmatrix} -1 & 0 \\ 1 & -1 \end{pmatrix}$, $C = \begin{pmatrix} -1 & 0 \\ 0 & -1 \end{pmatrix}$

(ii) $A = \begin{pmatrix} 1 & 2 & 3 \\ 0 & 1 & 2 \\ 0 & 0 & 1 \end{pmatrix}$, $B = \begin{pmatrix} 4 & 0 & 1 \\ 0 & 3 & 0 \\ 0 & 0 & 2 \end{pmatrix}$, $C = \begin{pmatrix} 2 & 0 & 1 \\ 0 & 2 & 1 \\ 1 & 0 & 2 \end{pmatrix}$.

APPENDIX TO CHAPTER 5

A proof of the Jordan form theorem

The Jordan form theorem (5.2.2) was stated as follows ;

Any matrix $A \in M_n\mathbb{C}$ is similar to a matrix J of the following form:

$$J = \begin{pmatrix} J_1(\lambda_1) & 0 & 0 & 0 \\ 0 & J_2(\lambda_2) & 0 & 0 \\ 0 & 0 & J_3(\lambda_3) & 0 \\ & & & \\ 0 & 0 & 0 & J_r(\lambda_r) \end{pmatrix}$$

where each $J_i(\lambda_i)$ is a Jordan block corresponding to an eigenvalue λ_i of A.

Each block $J_i(\lambda_i)$ is an $n_i x n_i$ matrix for some integer n_i.

The same eigenvalue λ_i may occur in more than one block, i.e. the λ_i need not all be different.

The matrix J is unique up to a permutation of the blocks $J_i(\lambda_i)$ and is known as the *Jordan canonical form* of A. (We will often refer to it simply as the *Jordan form* of A.)

The following information is useful in determining J ;

(i) The sum of the sizes of the blocks involving a particular eigenvalue of A equals the algebraic multiplicity of that eigenvalue

(ii) The number of blocks involving a particular eigenvalue of A equals the geometric multiplicity of that eigenvalue.

(iii) The largest block involving a particular eigenvalue of A equals the multiplicity of that eigenvalue as a root of the minimal polynomial of A.

Proof

To motivate our proof recall from (4.6.5) that an nxn matrix A is diagonalizable over $\mathbb{C}$ if and only if $\mathbb{C}^n$ has a basis consisting entirely of eigenvectors of A. The matrix P with these basis vectors as its columns gives a similarity transformation of A with a diagonal matrix whose entries are the eigenvalues of A. The Jordan form of a diagonalizable matrix consists entirely of $1{\times}1$ blocks. Motivated by (5.2.10) it is not difficult to see that an $n{\times}n$ matrix A is similar to a matrix in Jordan form if and only if $\mathbb{C}^n$ has a basis which can be partitioned into a collection of "strings" of vectors, a typical string consisting of vectors $v_1, v_2, \ldots, v_m$ for which $Av_1 = \lambda v_1$ and $Av_i = v_{i-1} + \lambda v_i$ for each $i = 2, 3, \ldots, m$. Each string corresponds to an $m{\times}m$ Jordan block involving the eigenvalue λ of A and the set of strings is in one-one correspondence with the set of blocks making up the Jordan form. To prove that A is similar to a matrix in Jordan form it suffices to produce a basis of the above kind.

We will prove by induction on n that any $n{\times}n$ matrix A is similar to a matrix in Jordan form. For $n = 1$ the result is true as any $1{\times}1$ matrix is already a $1{\times}1$ Jordan block.

Note first that it suffices to prove the result in the case when A is singular. To see this suppose that A is non-singular and let λ_0 be an eigenvalue of A. Then $A - \lambda_0 I$ is singular and if the matrix $P^{-1}(A - \lambda_0 I)P = J$ is a Jordan form then $P^{-1}AP = J + \lambda_0 I$ is also a Jordan form.

Now we let A be a singular $n{\times}n$ matrix and assume that any matrix of size less than n is similar to a Jordan form. We think of A as a linear operator on $\mathbb{C}^n$. The image of A, Im A, has dimension r where $r < n$ since A is singular. We can apply the induction hypothesis to the matrix representing the restriction of the linear operator A to the subspace Im A. This yields that Im A has a basis $v_1, v_2, \ldots, v_r$ which can be partitioned into strings in the manner described above.

Now let s be the dimension of (Ker A) $\cap$ (Im A), the intersection of the kernel and the image of A. Then in the basis $v_1, v_2, \ldots, v_r$ of Im A there must be s strings corresponding to $\lambda = 0$. If v is the last vector in one of these strings then we choose $w \in \mathbb{C}^n$ such that $Aw = v$. (This is possible as $v \in$ Im A !) This yields s new vectors $w_1, w_2, \ldots, w_s$ which we adjoin to the ends of the s strings. The corresponding Jordan blocks are thus increased in size by one.

Finally the part of Ker A complementary to Im A will have dimension n-r-s. We choose any basis $\{ x_1, x_2, \ldots, x_{n-r-s} \}$ for this complement Then $\{v_1, \ \ ,v_r, w_1, . \ , w_s, x_1, . \ , x_{n-r-s}\}$ is the desired basis of $\mathbb{C}^n$. We only need check that this set is linearly independent.

If $\sum_{i=1}^{r} \alpha_i v_1 + \sum_{i=1}^{s} \beta_i w_i + \sum_{i=1}^{n-r-s} \gamma_i x_i = 0$ for some scalars $\alpha_i, \beta_i, \gamma_i$ in $\mathbb{C}$ then applying A to this equation yields that $\sum_{i=1}^{r} \alpha_i A v_1 + \sum_{i=1}^{s} \beta_i A w_i = 0$.

Now each Aw_i equals some v_i which is the last vector in some string and this v_i cannot therefore appear with a coefficient α_i in front of it in the first part of the equation. Linear independence of $v_1, v_2, \ldots, v_r$ implies that each $\beta_i = 0$.

Our earlier equation can now be rewritten as follows ;

$$\sum_{i=1}^{r} \alpha_i v_1 = - \sum_{i=1}^{n-r-s} \gamma_i x_i .$$

Each side of this equation must be zero because the left-hand-side is in Im A while the right-hand-side is complementary to Im A. Linear independence of $v_1, \ldots, v_r$ and $x_1, \ldots, x_{n-r-s}$ implies that $\alpha_i = 0$ and $\gamma_i = 0$ for all i. This proves that any matrix is similar to a Jordan form.

We will now prove statements (i), (ii), and (iii). Since A and J are similar they have the same eigenvalues occurring with the same algebraic multiplicities by (4.3.2). Statement (i) follows at once because the eigenvalues of A are its diagonal entries. The eigenvalues of A and J have the same geometric multiplicities by (4.3.3). Statement (ii) follows from this and the easily verified fact that any Jordan block has a one-dimensional eigenspace. Statement (iii) follows from the observation that if J is an m×m Jordan block for λ then $J - \lambda I$ has index of nilpotency m, i.e. $J^m = 0$ but $J^k \neq 0$ for any $k < m$.

The final thing to be proved is that the Jordan form is unique apart from a permutation of the blocks. Let us suppose that A is similar to two different Jordan forms J_1 and J_2. Then there is some eigenvalue λ such that the blocks for λ in J_1 and J_2 differ in some way. Let k be the geometric multiplicity of λ as an eigenvalue of A. Let the blocks for λ in J_1 have sizes n_i with $n_1 \geq n_2 \geq \ldots \geq n_k$ and the blocks in J_2 have sizes m_i with $m_1 \geq m_2 \geq \ldots \geq m_k$. Then there is some i, $1 \leq i \leq k$, such that $n_i \neq m_i$ but $n_j = m_j$ for $j = 1, 2, \ldots, m - 1$. Without loss of generality we may assume $n_i > m_i$. Then $(J_1 - \lambda I)^{m_1} \neq 0$ whereas $(J_2 - \lambda I)^{m_1} = 0$. This is impossible since J_1 and J_2 are similar. Hence J_1 and J_2 can differ only by a permutation of the blocks and the proof is complete.

Comment

The above induction proof is due to the Russian mathematician Filippov. See Strang [St]. It is shorter than the traditional method of proof which involves an analysis of invariant subspaces of linear operators.

Chapter 6

SYMMETRIC AND HERMITIAN MATRICES

Nature is rich in symmetry and many of the matrices which arise from physical problems will have some kind of symmetry features. Usually these matrices will be real symmetric. The theory of real symmetric matrices can be regarded as a special case of the theory of complex hermitian matrices. We will examine various aspects and applications of the theory of real symmetric and complex hermitian matrices. We begin the chapter by discussing Schur's unitary triangularization theorem for complex matrices. We introduce the ideas of quadratic and hermitian forms and the way in which real symmetric and complex hermitian matrices arise from these forms. Topics which we treat include the notion of positive-definiteness for forms and matrices, the definition and calculation of the signature of a form, and the simultaneous reduction of a pair of forms. We go on to consider the eigenvalues of symmetric and hermitian matrices. The symmetry of these matrices makes available techniques different from those encountered earlier in this book In particular we consider the Rayleigh quotient from which estimates for the largest and smallest eigenvalues of a symmetric or hermitian matrix can be obtained, Rayleigh's principle which is used for estimating other eigenvalues, the Courant-Fisher min-max theorem and various applications.

6.1 **Schur triangular form**

6.1.1 **Triangular, unitary and orthogonal matrices**

We recall from chapter 1 the following definitions ;

The $n \times n$ matrix $A = (a_{ij})$ is *upper-triangular* if $a_{ij} = 0$ for all $i > j$, i.e. each entry below the main diagonal of A is zero.

The $n \times n$ matrix $A = (a_{ij})$ is *lower-triangular* if $a_{ij} = 0$ for all $i < j$, i.e. each entry of A above the main diagonal is zero.

The matrix $A \in M_n\mathbb{C}$ is *unitary* if $A^{-1} = \bar{A}^t$.

The matrix $A \in M_n\mathbb{R}$ is *orthogonal* if $A^{-1} = A^t$.

6.1.2 **Definition**

The matrices A and B in $M_n\mathbb{C}$ are *unitarily similar* if there exists a unitary matrix $P \in M_n\mathbb{C}$ such that $P^{-1}AP = B$.

The matrices A and B in $M_n\mathbb{R}$ are *orthogonally similar* if there exists an orthogonal matrix $P \in M_n\mathbb{R}$ such that $P^{-1}AP = B$.

6.1.3 **Schur's unitary triangularization theorem**

Any matrix $A \in M_n\mathbb{C}$ is unitarily similar to an upper triangular matrix, i.e. there exists $P \in M_n\mathbb{C}$ such that $P^{-1} = \bar{P}^t$ and $\bar{P}^tAP = T$ where T is an upper triangular matrix in $M_n\mathbb{C}$. If $A \in M_n\mathbb{R}$ and all the eigenvalues of A are real then the matrix P can be chosen to be a real orthogonal matrix, i.e. the matrix A is orthogonally similar to T.

Proof

Note first that the diagonal entries of T are necessarily the eigenvalues of A. This follows from (4.3.2) and (4.3.4).

The proof will be by induction on n. Any 1×1 matrix is triangular so that the result is true for $n = 1$. Now assume the theorem is true for all matrices of size less than n. Let $A \in M_n\mathbb{C}$ and let λ be an eigenvalue of A and v an eigenvector for λ with Euclidean norm one, i.e. $Av = \lambda v$ and $\| v \| = 1$.

Choose an orthonormal basis $\{w_1, w_2, \ldots, w_n\}$ of $\mathbb{C}^n$ with $w_1 = v$. (This is always possible since we may extend $\{v\}$ to a basis of $\mathbb{C}^n$ by (2.2.13) and then use the Gram-Schmidt process (2.7.7) to orthonormalize it.) All of this is with respect to the Euclidean norm on $\mathbb{C}^n$.

Let Q be the nxn matrix whose columns are $w_1, w_2, \ldots, w_n$. Q is unitary since $\bar{w}_i^t w_j = 0$ if $i \neq j$, $\bar{w}_i^t w_i = 1$ for all i, by the orthonormality of $\{w_1, w_2, \ldots, w_n\}$.

Remembering that $v = w_1$, $Av = \lambda v$, $\bar{v}^t v = 1$, $\bar{w}_i^t w_1 = 0$ for all $i \neq 1$, we see that for some $(n-1)\times(n-1)$ matrix C we can write $\bar{Q}^t AQ = \begin{pmatrix} \lambda & x \\ 0 & \\ \vdots & C \\ 0 & \end{pmatrix}$ where

$x = (\bar{v}^t A w_2 \; \bar{v}^t A w_3 \; \ldots \; \bar{v}^t A w_n)$, a row vector of length n-1.

If $n = 2$ the above matrix is upper triangular so the proof is finished in that case. For $n > 2$ we may assume, by the induction hypothesis, that C is unitarily similar to a triangular matrix, i.e. there exists a unitary $(n-1)\times(n-1)$ matrix V such that $V^{-1}CV$ is upper triangular.

Now let $U = \begin{pmatrix} 1 & 0 \cdots 0 \\ 0 & \\ \vdots & V \\ 0 & \end{pmatrix}$ and U will be unitary. Writing $P = QU$ it is easy to see that P is unitary and that $\bar{P}^t AP = \bar{U}^t \bar{Q}^t AQU$ is of the form $\begin{pmatrix} \lambda & * & * & * \\ 0 & & & \\ \vdots & & V^{-1}CV & \\ 0 & & & \end{pmatrix}$

which is upper triangular. This completes the proof by induction.

If $A \in M_n\mathbb{R}$ and all the eigenvalues of A are real then the above procedure can be done using real orthogonal matrices instead of unitary matrices so that P may be chosen to be real orthogonal.

6.1.4 **Exercise**

If A is a hermitian matrix, i.e. $A = \bar{A}^t$, show that A is unitarily similar to a diagonal matrix. (Show that the matrix T in (6.1.3) is necessarily diagonal!)

6.1.5 **Comment**

The triangular matrix T obtained in 6.1.3 is not unique and neither is the unitary matrix P. The matrix T is often referred to as a *Schur canonical form* for the matrix A. In view of the lack of uniqueness of T it is somewhat misleading to use the phrase *canonical form* but the terminology is widely used.

6.1.6 **Corollary**

Any matrix $A \in M_n\mathbb{C}$ is unitarily similar to a lower triangular matrix, i.e. there exists $P \in M_n\mathbb{C}$ such that $P^{-1} = \bar{P}^t$ and $\bar{P}^tAP = T$ which is lower triangular. If $A \in M_n\mathbb{R}$ and has all its eigenvalues real then P may be chosen to be real orthogonal, i.e. A is orthogonally similar to T.

Proof

Applying 6.1.3 to $\bar{A}^t$ instead of A we find that there exists a matrix $Q \in M_n\mathbb{C}$ with $Q^{-1} = \bar{Q}^t$ and $\bar{Q}^t\bar{A}^tQ = T$, T being upper triangular. Taking the conjugate transpose of this last equation yields that $\bar{Q}^tAQ = \bar{T}^t$ and since $\bar{T}^t$ is lower triangular this completes the proof.

6.1.7 **Example**

Find a Schur canonical form for the matrix $A = \begin{pmatrix} 2 & 1 & 0 \\ 2 & 3 & 0 \\ -1 & -1 & 1 \end{pmatrix}$.

A straightforward calculation shows that the characteristic polynomial of A is $(1-\lambda)^2(4-\lambda)$. The eigenvalues of A are 1, 1, 4 so that in particular they are all real. Hence by (6.1.3) there exists a real orthogonal matrix P such that P^tAP is upper triangular.

Proceeding as in the proof of (6.1.3) we construct a unitary matrix Q whose first column is an eigenvector of A with norm 1.

We take the eigenvalue $\lambda = 1$ and the corresponding eigenvector $\begin{pmatrix} 0 \\ 0 \\ 1 \end{pmatrix}$ as column 1 of Q. We then fill in the other two columns of Q in the easiest way possible that will make Q orthogonal, i.e. that will make the columns of Q orthonormal.

Thus we may take $Q = \begin{pmatrix} 0 & 0 & 1 \\ 0 & 1 & 0 \\ 1 & 0 & 0 \end{pmatrix}$ and then an easy calculation shows that

$$Q^tAQ = \begin{pmatrix} 1 & -1 & -1 \\ 0 & 3 & 2 \\ 0 & 1 & 2 \end{pmatrix}.$$

We now confine ourselves to the 2x2 matrix $\begin{pmatrix} 3 & 2 \\ 1 & 2 \end{pmatrix}$ which also has $\lambda = 1$ as an eigenvalue. Taking $\begin{pmatrix} 1/\sqrt{2} \\ -1/\sqrt{2} \end{pmatrix}$ as an eigenvector of norm 1 for $\lambda = 1$ we may take

$V = \begin{pmatrix} 1/\sqrt{2} & 1/\sqrt{2} \\ -1/\sqrt{2} & 1/\sqrt{2} \end{pmatrix}$ and then take $U = \begin{pmatrix} 1 & 0 & 0 \\ 0 & 1/\sqrt{2} & 1/\sqrt{2} \\ 0 & -1/\sqrt{2} & 1/\sqrt{2} \end{pmatrix}$. The matrix

$P = QU = \begin{pmatrix} 0 & -1/\sqrt{2} & 1/\sqrt{2} \\ 0 & 1/\sqrt{2} & 1/\sqrt{2} \\ 1 & 0 & 0 \end{pmatrix}$ and by calculation we see that $P^tAP = \begin{pmatrix} 1 & 0 & -\sqrt{2} \\ 0 & 1 & 1 \\ 0 & 0 & 4 \end{pmatrix}$

which is a Schur canonical form for A.

Problems 6A

1. Let $B \in M_n\mathbb{C}$ and let B be skew-hermitian, i.e. $\bar{B}^t = -B$.

(i) Show that all of the eigenvalues of B are of the form $i\alpha$ for some $\alpha \in \mathbb{R}$.

(ii) Show that the matrix exponential e^B is unitary.

2. Let $B \in M_n\mathbb{C}$ and suppose that $I + B$ is invertible.

Let $\gamma(B) = (I + B)^{-1}(I - B)$. If $\bar{B}^t = -B$ show that $\gamma(B)$ is unitary. If B is unitary show that $\gamma(B)$ is skew-hermitian, i.e. $\overline{\gamma(B)}^t = -\gamma(B)$

(The map γ is known as the *Cayley transform.*)

3. Let $A = \begin{pmatrix} 1 & 2 \\ -1 & -2 \end{pmatrix}$, $B = \begin{pmatrix} -1 & 0 \\ 0 & 0 \end{pmatrix}$, $C = \begin{pmatrix} 0 & -2 \\ 0 & -1 \end{pmatrix}$. Find an orthogonal matrix P such that $P^tAP = B$ and an orthogonal matrix Q such that $Q^tAQ = C$, i.e. B and C are each Schur canonical forms for A.

4. Show that $\begin{pmatrix} 0 & 1 \\ 0 & 7 \end{pmatrix}$ is a Schur canonical form for $\begin{pmatrix} 1 & 2 \\ 3 & 6 \end{pmatrix}$.

Why can $\begin{pmatrix} 0 & 2 \\ 0 & 7 \end{pmatrix}$ not be a Schur canonical form for $\begin{pmatrix} 1 & 2 \\ 3 & 6 \end{pmatrix}$?

5. Let $A \in M_n\mathbb{R}$. Show that there exists an orthogonal matrix P such that P^tAP is a block upper triangular matrix of the form

$$\begin{pmatrix} T_1 & * & * & * \\ 0 & T_2 & & * \\ & & T_3 & \\ 0 & & & * \\ 0 & & & T_r \end{pmatrix}$$

where each block T_i, $i = 1,2,\ldots,r$, is either a 1x1 block or a 2x2 block with a pair of complex conjugate eigenvalues.

(The above block upper triangular matrix is known as a real Schur canonical form.)

(*Hint* - modify the proof of (6.1.3) as follows ;

If $\lambda = a + ib$ is an eigenvalue with $b \neq 0$ and $Av = \lambda v$ for $v \in \mathbb{C}^n$ then write $v = x + iy$ for vectors $x, y \in \mathbb{R}^n$. Show that $\{x,y\}$ is a linearly independent set in $\mathbb{R}^n$ which can be orthonormalized into a set $\{w_1,w_2\}$ by the Gram-Schmidt process. Let Q be an orthogonal matrix with w_1 and w_2 as its first two columns. Show that Q^tAQ is of the form $\begin{pmatrix} \mathbf{T} & * \ * & * \ * \\ & * \ * & * \ * \\ 0 \ 0 & & \\ 0 \ 0 & \mathbf{B} & \\ & & \\ 0 \ 0 & & \end{pmatrix}$ where $\mathbf{B}$ is an $(n - 2)x(n - 2)$-matrix and $\mathbf{T}$ is a $2x2$ matrix which is similar to $\begin{pmatrix} a & b \\ -b & a \end{pmatrix}$. Then proceed by induction.)

6. Let $\sigma(A)$ denote the spectrum of the nxn matrix A, i.e. the set of all eigenvalues of A.

Show that $\sigma(A^k) = \{\lambda^k : \lambda \in \sigma(A)\}$ for each natural number k.

Show that $\sigma(p(A)) = \{ p(\lambda) : \lambda \in \sigma(A) \}$ for each polynomial p in one variable.

(*Hint* - consider a Schur canonical form for A.)

7. Let A be an mxm matrix and B an nxn matrix. The *Kronecker product* of A with B, denoted by $A \otimes B$, is the $mnxmn$ matrix made up of blocks as follows ;

$$A \otimes B = \begin{pmatrix} a_{11}B & a_{12}B & \dots & a_{1m}B \\ a_{21}B & a_{22}B & \dots & a_{2m}B \\ \cdot & & & \\ \cdot & & & \\ a_{m1}B & a_{m2}B & \dots & a_{mm}B \end{pmatrix}$$

(i) If A_1, A_2 are mxm matrices, and B_1, B_2 are nxn matrices show that $(A_1 \otimes B_1)(A_2 \otimes B_2) = A_1A_2 \otimes B_1B_2$.

(ii) Show that $(A \otimes B)^t = A^t \otimes B^t$.

(iii) Show that if A and B are unitary matrices then $A \otimes B$ is unitary.

(iv) Show that if A and B are upper triangular matrices then $A \otimes B$ is upper triangular.

(v) Use Schur canonical forms for A and B to deduce that the spectrum

$\sigma(A \otimes B) = \{ \lambda_i\mu_j : \lambda_i \in \sigma(A), \mu_j \in \sigma(B) \}$.

(vi) Deduce the following by using (4.3.5) and the above ;

$$\text{trace } (A \otimes B) = (\text{trace } A)(\text{trace } B)$$

$$\det (A \otimes B) = (\det A)^n (\det B)^m$$

(The Kronecker product of two rectangular matrices can be defined in similar fashion. The Kronecker product can be formulated in a more conceptual framework via the notion of the tensor product of vector spaces. This viewpoint is beyond the scope of this book.)

6.2 Real quadratic forms

6.2.1 Definition

Let V be a finite-dimensional vector space over $\mathbb{R}$.

A *bilinear form* on V is a mapping $\phi : V \times V \longrightarrow \mathbb{R}$ which is linear in each of the two variables, i.e. ϕ satisfies the following;

$\phi(v_1 + v_2, w) = \phi(v_1, w) + \phi(v_2, w)$ for all v_1, v_2, w in V,

$\phi(\alpha v, w) = \alpha\phi(v, w)$ for all $\alpha \in \mathbb{R}$, for all v, w in V,

$\phi(v, w_1 + w_2) = \phi(v, w_1) + \phi(v, w_2)$ for all v, w_1, w_2 in V,

$\phi(v, \alpha w) = \alpha\phi(v, w)$ for all $\alpha \in \mathbb{R}$, for all v, w in V.

A *symmetric bilinear form* on V is a bilinear form on V with the extra property that $\phi(v, w) = \phi(w, v)$ for all v, w in V.

6.2.2 Definition

Let V be a finite-dimensional vector space over $\mathbb{R}$.

A *quadratic form* on V is a mapping $q : V \longrightarrow \mathbb{R}$ satisfying the following two conditions;

(i) $q(\alpha v) = \alpha^2 q(v)$ for all $\alpha \in \mathbb{R}$, for all $v \in V$,

(ii) the mapping $\psi : V \times V \longrightarrow \mathbb{R}$, $\psi(v, w) = q(v + w) - q(v) - q(w)$ is a bilinear form on V.

6.2.3 **Proposition**

There is a one- one correspondence between the set of all quadratic forms on the real vector space V and the set of all symmetric bilinear forms on V.

Proof

Let $\phi: V \times V \longrightarrow \mathbb{R}$ be a symmetric bilinear form on V. Defining $q: V \longrightarrow \mathbb{R}$ by $q(v) = \phi(v,v)$ for each $v \in V$ yields a quadratic form on V.

Conversely, given a quadratic form $q: V \longrightarrow \mathbb{R}$, defining a mapping $\phi: V \times V \longrightarrow \mathbb{R}$ by $\phi(v,w) = (1/2)\{q(v + w) - q(v) - q(w)\}$ for each v,w in V yields a symmetric bilinear form on V. It is straightforward to check that the above gives a one-one correspondence between the set of all quadratic forms on V and the set of all symmetric bilinear forms on V.

(Note that the factor 1/2 is necessary to ensure a one-one correspondence.)

6.2.4 **Remark**

Proposition 6.2.3 shows that the concepts of "real quadratic form" and "real symmetric bilinear form" are interchangeable. We will pass from one concept to the other at our convenience.

6.2.5 **Example**

Let $V = \mathbb{R}^n$ and let $A \in M_n\mathbb{R}$, $x \in \mathbb{R}^n$, $y \in \mathbb{R}^n$.

Then the mapping $\phi: V \times V \longrightarrow \mathbb{R}$ given by $\phi(x,y) = x^tAy$ is a bilinear form on V. If A is a symmetric matrix then ϕ is a symmetric bilinear form.

The quadratic form corresponding to ϕ is the map $q: V \longrightarrow \mathbb{R}$ given by $q(x) = x^tAx$.

Note that any inner product on a real vector space, as defined in (2.7), is a symmetric bilinear form with the extra property of being positive-definite.

We will see soon that any quadratic form or symmetric bilinear form on a finite-dimensional vector space V is expressible, via some symmetric matrix A, in the manner of the above example.

Note that if we write $x = (x_i)$ and $A = (a_{ij})$ then we may regard $q(x) = x^tAx$ as a function of n real variables $x_1,x_2,\ldots,x_n$.
Specifically $q(x_1,x_2,\ldots,x_n) = \sum_i \sum_j a_{ij}x_ix_j$.
In the language of functions the quadratic form is a homogeneous polynomial of degree two in the n variables $x_1,x_2,\ldots,x_n$.

6.2.6 **Example**

Any sum of squares $x_1^2 + x_2^2 + \ldots + x_n^2$ may be regarded as a quadratic form. We simply put $A = I$, the identity matrix, in example (6.2.5). Thus quadratic forms can be viewed as a generalization of sums of squares.

6.2.7 **Geometric examples**

In (4.7) we discussed conic sections and quadric surfaces. These can be described in the language of quadratic forms.

Let $V = \mathbb{R}^2$ and let a,b,c be fixed real numbers. Let $q:\mathbb{R}^2 \longrightarrow \mathbb{R}$ be defined by $q(x,y) = ax^2 + 2bxy + cy^2$. In matrix notation this can be written $q(x,y) = (x\ y)\begin{pmatrix} a & b \\ b & c \end{pmatrix}\begin{pmatrix} x \\ y \end{pmatrix}$.

The curve $q(x,y)$ = constant will represent a conic section, i.e. ellipse, parabola, hyperbola, etc., the particular kind of conic section depending on the values of a,b, and c.

Let $V = \mathbb{R}^3$ and let a,b,c,h,g,f be fixed real numbers. Let $q:\mathbb{R}^3 \longrightarrow \mathbb{R}$ be defined by $q(x,y,z) = ax^2 + by^2 + cz^2 + 2hxy + 2gxz + 2fyz$. In matrix notation $q(x,y,z) = (x\ y\ z)\begin{pmatrix} a & h & g \\ h & b & f \\ g & f & c \end{pmatrix}\begin{pmatrix} x \\ y \\ z \end{pmatrix}$.

The surface $q(x,y,z)$ = constant will represent a quadric surface in $\mathbb{R}^3$, i.e. ellipsoid, hyperboloid, paraboloid etc., the particular kind of surface depending on a,b,c,h,g,f.

6.2.8 **Proposition**

Let V be an n-dimensional real vector space. The following three sets are in one-one correspondence with each other;

(i) the set of all quadratic forms on V.

(ii) the set of all symmetric bilinear forms on V.

(iii) the set of all real symmetric n×n matrices.

Proof

We have seen already in (6.2.3) that (i) and (ii) are in one-one correspondence. We will now exhibit a one-one correspondence between (ii) and (iii).

Choosing a basis $\{v_1,v_2,...,v_n\}$ of V we associate to the symmetric bilinear form $\phi:V\times V \longrightarrow \mathbb{R}$ the real symmetric matrix with entries $\phi(v_i,v_j)$. Conversely let $A = (a_{ij})$ be a real symmetric n×n matrix . Define a form $\phi:V\times V \longrightarrow \mathbb{R}$ by $\phi(v_i,v_j) = a_{ij}$ for the basis vectors $v_1,v_2,...,v_n$ and then extend by linearity, i.e. define $\phi(\sum_i \alpha_i v_i, \sum_j \beta_j v_j) = \sum_i \sum_j \alpha_i \beta_j a_{ij}$ for all the possible choices of $\alpha_1,\alpha_2,...,\alpha_n,\beta_1,\beta_2,...,\beta_n$ in $\mathbb{R}$. Then ϕ is a symmetric bilinear form. It is straightforward to check that the above yields a one-one correspondence between (ii) and (iii).

6.2.9 **Comment**

The correspondence between (ii) and (iii) in (6.2.8) depends on the choice of basis $\{v_1,v_2,...,v_n\}$ of V. A different choice of basis, say $\{w_1,w_2,...,w_n\}$, will yield a different one-one correspondence between (ii) and (iii). The matrix corresponding to ϕ for the basis $\{v_1,v_2,...,v_n\}$ will differ from that for the basis $\{w_1,w_2,...,w_n\}$. The relationship between these two matrices is given by the following;

6.2.10 Lemma

Let $\phi : V\times V \longrightarrow \mathbb{R}$ be a symmetric bilinear form. Let A and B be the matrices with entries $\phi(v_i,v_j)$ and $\phi(w_i,w_j)$ respectively for the two bases $\{v_1,v_2,\ldots,v_n\}$ and $\{w_1,w_2,\ldots,w_n\}$ of V. Then $P^tAP = B$ where P is the matrix of change of basis from $\{v_1,v_2,\ldots,v_n\}$ to $\{w_1,w_2,\ldots,w_n\}$.

Proof

The entries of P are obtained by expressing each w_i as a linear combination of $v_1,v_2,\ldots,v_n$. If P has entries p_{ij} then we have $w_j = \sum_{i=1}^{n} p_{ij}v_i$ for each $j = 1,2,\ldots,n$. The matrix B will have entries $\phi(w_i,w_j)$ where $\phi(w_i,w_j) = \phi(\sum_{k=1}^{n} p_{ki}v_k, \sum_{m=1}^{n} p_{mj}v_m) = \sum_{k=1}^{n}\sum_{m=1}^{n} p_{ki}\phi(v_k,v_m)p_{mj}$ because of the bilinearity of ϕ. Since A has entries $\phi(v_i,v_j)$ it is a straightforward exercise to check that, by the rules of matrix multiplication, $B = P^tAP$.

6.2.11 Definition

Two real symmetric $n\times n$ matrices A and B are *congruent* if there exists invertible $n\times n$ matrix P such that $P^tAP = B$.

6.2.12 Remark

Lemma (6.2.10) thus says that the matrices representing a given symmetric bilinear form with respect to two different bases must be congruent. The matrix P in (6.2.10) is invertible because it is a matrix of change of basis.

6.2.13 Further examples

We saw in (6.2.7) how quadratic forms arise in connection with conic sections and quadric surfaces. We now describe briefly a few more situations where symmetric matrices and quadratic forms crop up naturally.

(i) Calculus of several variables

The Hessian matrix $(\partial^2 f/\partial x_i \partial x_j)$ of a twice continuously differentiable function f of n real variables $x_1, x_2, \ldots, x_n$ is a symmetric matrix. An analysis of this matrix and its corresponding quadratic form at a critical point of f enables one to determine the nature of the critical point. See also Example (c) of (1.7) and Example (a) of (4.7).

(ii) Statistics

The covariance matrix for a set of random variables $X_1, X_2, \ldots, X_n$ has entries $E((X_i - \mu_i)(X_j - \mu_j))$ where μ_i is the mean of the X_i and E denotes the expectation. This symmetric matrix is used in statistical analysis. See also (1.7) Example (h).

(iii) Mechanics

Physical quantities such as energy, inertia, momentum can often be approximated by quadratic forms.

(iv) Theory of graphs and networks

The adjacency matrix of a graph with n nodes has entry $a_{ij} = 1$ if the i-th and j-th nodes are joined by an edge of the graph and $a_{ij} = 0$ otherwise. For an undirected graph this matrix is symmetric. See also (1.7), Example (f).

6.3 **Hermitian forms**

The notions of symmetric bilinear form and quadratic form are meaningful if we replace the real numbers $\mathbb{R}$ by the complex numbers $\mathbb{C}$ in definitions (6.2.)1 and (6.2.2). However complex symmetric bilinear forms and complex quadratic forms do not have such useful properties as in the real case and do not occur so often in a natural way. The appropriate generalization to $\mathbb{C}$ is that of a hermitian form.

6.3.1 **Definition**

Let V be a finite-dimensional complex vector space. A *hermitian form* on V is a mapping $h:V\times V \longrightarrow \mathbb{C}$ satisfying the following two conditions ;

(i) h is sesquilinear, i.e. h satisfies the following ;

$h(v_1 + v_2,w) = h(v_1,w) + h(v_2,w)$ for all v_1, v_2, w in V.

$h(\alpha v,w) = \bar{\alpha}h(v,w)$ for all $\alpha \in \mathbb{C}$, for all v, w in V.

$h(v,w_1 + w_2) = h(v,w_1) + h(v,w_2)$ for all v_1, v_2, w in V.

$h(v,\alpha w) = \alpha h(v,w)$ for all $\alpha \in \mathbb{C}$, for all v, w in V.

(ii) h is hermitian symmetric

i.e. $h(v,w) = \overline{h(w,v)}$ for all v,w in V.

6.3.2 **Remark**

The word *sesquilinear* means "one and a half times linear". The hermitian form h fails to be bilinear only because multiplying the first variable by a scalar requires conjugation of that scalar when it is taken outside of h. We will describe this last feature of h by saying that h is anti-linear in the first variable and linear in the second variable.

Some authors adopt a different convention from us and define a hermitian form to be linear in the first variable and anti-linear in the second variable. The theory of hermitian forms is the same whichever convention you adopt.

6.3.3 **Proposition**

Let V be an n-dimensional complex vector space. The set of all hermitian forms on V is in one-one correspondence with the set of all complex hermitian $n\times n$ matrices.

Proof

The proof is essentially the same as that for the correspondence of (ii) and (iii) of (6.2.8), i.e. we choose a basis $\{v_1,v_2,\ldots,v_n\}$ of V and take the hermitian matrix with entries $h(v_i,v_j)$.

6.3.4 **Lemma**

Let $h:V\times V \longrightarrow \mathbb{C}$ be a hermitian form. Let A and B be the matrices with entries $h(v_i,v_j)$ and $h(w_i,w_j)$ respectively for the two bases $\{v_1,v_2,\ldots,v_n\}$ and $\{w_1,w_2,\ldots,w_n\}$ of V. Then $\bar{P}^tAP = B$ where P is the matrix of change of basis from $\{v_1,\ldots,v_n\}$ to $\{w_1,\ldots,w_n\}$.

Proof

Similar to the proof of (6.2.10), the anti-linearity in the first variable resulting in the appearance of $\bar{P}^t$ rather than P^t as in (6.2.10).

6.3.5 **Definition**

Two complex hermitian matrices A and B are *congruent* if there exists an invertible $n\times n$ matrix P such that $\bar{P}^tAP = B$.

6.3.6 **Remark**

Lemma (6.3.4) says that the matrices representing a hermitian form with respect to two different bases are congruent.

6.3.7. **Comment**

We have in fact already met some examples of hermitian forms earlier in this book. An inner product on a complex vector space, as defined in (2.7), is a hermitian form with the extra property of being positive-definite. In particular the Euclidean inner product $h:\mathbb{C}^n\times\mathbb{C}^n \longrightarrow \mathbb{C}$, $h(v,w) = \bar{v}^tw$ is a hermitian form. More generally if A is any $n\times n$ hermitian matrix then $h:\mathbb{C}^n\times\mathbb{C}^n \longrightarrow \mathbb{C}$, $h(v,w) = \bar{v}^tAw$, is a hermitian form.

Problems 6B

1. Let $\phi_i:V_i\times V_i \longrightarrow \mathbb{R}$, $i = 1,2$,be symmetric bilinear forms. The forms ϕ_1 and ϕ_2 are said to be *equivalent* if there exists a vector space isomorphism $\gamma:V_1 \longrightarrow V_2$ such that $\phi_2(\gamma(x),\gamma(y)) = \phi_1(x,y)$ for all x,y in V_1.

Two quadratic forms $q_i:V_i \longrightarrow \mathbb{R}$, $i = 1,2$,are said to be *equivalent* if there exists a vector space isomorphism $\gamma:V_1 \longrightarrow V_2$ such that $q_2(\gamma(x)) = q_1(x)$ for all $x \in V_1$.

Prove that two symmetric bilinear forms are equivalent if and only if the corresponding quadratic forms are equivalent, the correspondence being as in (6.2.3).

2. Show that two real symmetric bilinear forms are equivalent if and only if any pair of matrices representing the two forms are congruent.

3. Let $h:V \times V \longrightarrow \mathbb{C}$ be a complex hermitian form. Viewing V as a real vector space, (whose dimension will be double its dimension as a complex vector space), show that $q:V \longrightarrow \mathbb{R}$, $q(v) = h(v,v)$, is a real quadratic form. This form q is known as *the underlying quadratic form* of h.

If h is represented by the hermitian matrix A with respect to a basis $\{v_1,v_2,\ldots,v_n\}$ of V show that $A = B + iC$ where B and C are real $n \times n$ matrices with $B^t = B$ and $C^t = -C$.

Show that $\{v_1,v_2,\ldots,v_n,iv_1,iv_2,\ldots,iv_n\}$ is a basis for V as a real vector space and that, with respect to this basis, the underlying quadratic form of h is represented by the matrix $\begin{pmatrix} B & -C \\ C & B \end{pmatrix}$.

4. Let $\phi:V \times V \longrightarrow \mathbb{R}$ be a real symmetric bilinear form. Show that if $\phi(v,v) = 0$ for all $v \in V$ then $\phi(v,w) = 0$ for all v,w in V.

[*Hint* - consider $\phi(v + w,v + w)$]

5. Let $h:V \times V \longrightarrow \mathbb{C}$ be a complex hermitian form. Show that if $h(v,v) = 0$ for all $v \in V$ then $h(v,w) = 0$ for all $v,w \in V$.

[*Hint* - by considering $h(v + w,v + w)$ deduce that $h(v,w)$ has zero real part. Then use the fact that the real part of $h(iv,w)$ equals the imaginary part of $h(v,w)$].

6.4 Reduction to diagonal form

From now on in this chapter we will state and prove everything for hermitian forms and matrices. All of our results will hold also for quadratic forms and real symmetric matrices since these can be viewed as special cases of hermitian forms and matrices. We simply ignore the complex conjugation sign whenever it appears because a real number equals its own conjugate.

We saw in (4.3.6) that all of the eigenvalues of a hermitian matrix are real. This fact can also be deduced from Schur's theorem since any Schur canonical form T for a hermitian matrix must satisfy $T = \bar{T}^t$, i.e. T is diagonal with real entries. Indeed Schur's theorem tells us the following;

6.4.1 Proposition

Any hermitian matrix A is congruent to a real diagonal matrix.

Proof

By Schur's theorem $\bar{P}^t AP = T$ where P is unitary and T is upper triangular. Now $A = \bar{A}^t$ implies that $T = \bar{T}^t$ so that T is diagonal with real entries.

6.4.2 Corollary

Let $h: V \times V \longrightarrow \mathbb{C}$ be a hermitian form on a finite-dimensional complex vector space V. There exists a basis of V with respect to which h has a real diagonal matrix representation.

Proof

Let $\{v_1, v_2, \ldots, v_n\}$ be any basis of V and A be the hermitian matrix with entries $h(v_i, v_j)$. Using the notation in (6.4.1) we see that $\bar{P}^t AP = T$ with T real diagonal and T is the matrix of h with respect to the basis $\{Pv_1, Pv_2, \ldots, Pv_n\}$.

6.5 Positive-definite forms and matrices

6.5.1 Definition

The hermitian form $h:V\times V \longrightarrow \mathbb{C}$ is *positive-definite* provided that $h(v,v) > 0$ for all non-zero vectors $v \in V$.

The hermitian $n\times n$ matrix A is *positive-definite* provided that $\bar{v}^t A v > 0$ for all non-zero vectors $v \in \mathbb{C}^n$.

6.5.2 Exercise

Show that the hermitian form h is positive-definite if and only if any matrix representing h is positive-definite. [*Hint* - show that A is positive-definite if and only if $\bar{P}^t A P$ is positive-definite for each invertible matrix P.]

6.5.3 Comment

Definition (6.5.1) applies to real symmetric bilinear forms and real symmetric matrices. A real quadratic form $q:V \longrightarrow \mathbb{R}$ is said to be *positive-definite* if and only if the symmetric bilinear form corresponding to q is positive-definite.

(Thus q is positive-definite if and only if $q(v) > 0$ for all non-zero vectors $v \in V$.)

6.5.4 Principal minors

Let $A = (a_{ij})$ be an $n\times n$ hermitian matrix.

The *principal minors* of the matrix A are the values $D_1, D_2, \ldots, D_n$ defined as follows;

$$D_1 = a_{11},\ D_2 = \det\begin{pmatrix} a_{11} & a_{12} \\ a_{21} & a_{22} \end{pmatrix},\ D_3 = \det\begin{pmatrix} a_{11} & a_{12} & a_{13} \\ a_{21} & a_{22} & a_{23} \\ a_{31} & a_{32} & a_{33} \end{pmatrix}, \ldots, D_n = \det A$$

i.e. for each $k = 1,2,\ldots,n$ the number D_k is the determinant of the top left-hand corner $k\times k$ block of the matrix A.

Note that each D_k is a real number because it is the determinant of a hermitian matrix. ($A = \bar{A}^t$ implies $\det A = \det \bar{A}^t = \det \bar{A}$ by part (2) of (1.4.2) and $\det \bar{A} = \overline{\det A}$ for any matrix A.)

The following is a famous criterion for positive-definiteness;

6.5.5 **Proposition** (Principal minors criterion)

The hermitian matrix A is positive-definite if and only if each of the principal minors of A is positive.

Proof

Note first that if A is positive-definite then all of the eigenvalues of A must be positive. (If $Ax = \lambda x$ for $\lambda < 0$ and $x \neq 0$, $x \in \mathbb{C}^n$, then $\bar{x}^t Ax = \lambda \bar{x}^t x < 0$). Hence $D_n = \det A$ is positive since by (4.3.5) it is the product of all the eigenvalues of A. Also if A is positive-definite and $h: \mathbb{C}^n \times \mathbb{C}^n \longrightarrow \mathbb{C}$ is given by $h(x,y) = \bar{x}^t Ay$ then the restriction of h to the k-dimensional subspace of $\mathbb{C}^n$ spanned by the first k vectors of the standard basis of $\mathbb{C}^n$ must be positive-definite. Hence $D_k > 0$ for each k since D_k is the determinant of this restriction of h. This proves the implication one way around.

We prove the converse implication by induction on n. For $n = 1$ the result is clearly true so assume it is true for all $(n-1)\times(n-1)$ hermitian matrices. Let $A = \bar{A}^t$ be an $n \times n$ matrix and write $h(x,x) = \bar{x}^t Ax = \sum_i \sum_j a_{ij} \bar{x}_i x_j$ where $x = (x_i) \in \mathbb{C}^n$.

It is straightforward but laborious to check that we may write $h(x,x) = a_{11}\bar{y}y - a_{11}\bar{z}z + \sum_{i=2}^{n} \sum_{j=2}^{n} a_{ij}\bar{x}_i x_j$ where y and z are defined by

$$y = x_1 + (a_{12}/a_{11})x_2 + (a_{13}/a_{11})x_3 + \ldots\ldots + (a_{1n}/a_{11})x_n\ , \quad z = y - x_1.$$

Since z depends only on $x_2, x_3, \ldots, x_n$ we can define the vector $x' = (x_2, x_3, \ldots, x_n)$ and the function h' of n-1 variables by

$$h'(x',x') = \sum_{i=2}^{n} \sum_{j=2}^{n} a_{ij}\bar{x}_i x_j - a_{11}\bar{z}z.$$

Now when we write $b_{ij} = a_{ij} - \{(a_{i1}a_{1j})/a_{11}\}$, we will find that $h'(x',x') = \sum_{i=2}^{n} \sum_{j=2}^{n} b_{ij}x_i x_j$. The matrix $B = (b_{ij})$ is an $(n-1)\times(n-1)$ hermitian matrix since A is hermitian. (Note that in our notation the entries of B are b_{ij} where i and j range from 2 to n.)

The key observation now is that the principal minors D_k of A satisfy

$$D_k = \det \begin{pmatrix} a_{11} & 0 & 0 & & 0 \\ 0 & b_{22} & b_{23} & & b_{2k} \\ 0 & b_{32} & b_{33} & & b_{3k} \\ & & & & \\ 0 & b_{k2} & b_{k3} & & b_{kk} \end{pmatrix}.$$

(Notice that the way in which the b_{ij} are constructed from the a_{ij} is precisely what is needed to reduce D_k by elementary row and column operations to a matrix with zero everywhere in the first row and column except for the entry a_{11}.) Since all of the D_k are positive it follows that all of the principal minors of B are positive. Hence B is positive-definite by the inductive assumption. This implies that $h'(x',x') > 0$ for all $x' \neq 0$ and so $h(x,x) > 0$ for all $x \neq 0$. This completes the proof that A is positive-definite.

6.5.6 **Example**

$$A = \begin{pmatrix} 1 & 2 & 0 \\ 2 & 6 & 1 \\ 0 & 1 & 2 \end{pmatrix}$$

$D_1 = 1$, $D_2 = 2$. $D_3 = 3$, so that A is positive-definite by (6.5.5).

6.5.7 **Example**

$$A = \begin{pmatrix} 1 & i & 0 & 1 \\ -i & 1 & 2 & 0 \\ 0 & 2 & 1 & 0 \\ 1 & 0 & 0 & 1 \end{pmatrix}$$

$D_1 = 1$, $D_2 = 0$, so that A is not positive-definite by (6.5.5).

6.5.8 **Remark**

See Problems 6C for more about positive-definiteness.

6.5.9 **Definition**

The hermitian form $h:V \times V \longrightarrow \mathbb{C}$ is *negative-definite* provided that $h(v,v) < 0$ for all non-zero vectors $v \in V$.

The hermitian $n \times n$ matrix A is *negative-definite* provided that $\bar{v}^t A v < 0$ for all non-zero vectors $v \in \mathbb{C}^n$.

6.5.10 **Note**

The hermitian form h will be negative-definite if and only if the form $-h$ is positive-definite. Similarly a hermitian matrix A is negative-definite if and only if $-A$ is positive-definite. See problem 7 of Problems 6C for a principal minors criterion for negative-definiteness.

6.5.11 **Definition**

The hermitian form $h:V \times V \longrightarrow \mathbb{C}$ is *non-singular* provided that any matrix representing h is non-singular.

(This definition does not depend on the particular matrix chosen to represent h since $\det \bar{P}^t AP \neq 0$ if and only if $\det A \neq 0$ for any non-singular matrix P).

6.5.12 **Exercise**

Show that any positive-definite or negative-definite hermitian form is non-singular. By considering the matrix $\begin{pmatrix} 1 & 0 \\ 0 & -1 \end{pmatrix}$ show that the converse of this last statement is false.

Problems 6C

1. Show that a hermitian form $h:V \times V \longrightarrow \mathbb{C}$ is non-singular if and only if $h(v,w) = 0$ for all $w \in V$ implies that $v = 0$.

2. If $A = (a_{ij})$ is a positive-definite hermitian matrix show that

(i) $a_{ii} > 0$ for all i,

(ii) $a_{ii}a_{jj} > |a_{ij}|^2$ for all $i \neq j$.

3. Determine whether or not each of the following matrices is positive-definite;

$$\text{(i)} \begin{pmatrix} 2 & -2 & 1 \\ -2 & 1 & 0 \\ 1 & 0 & 4 \end{pmatrix}, \text{(ii)} \begin{pmatrix} 1 & -1 & 2 & 0 \\ -1 & 2 & -1 & 1 \\ 2 & -1 & 6 & 2 \\ 0 & 1 & 2 & 3 \end{pmatrix}, \text{(iii)} \begin{pmatrix} 1 & -1 & 2 & 0 \\ -1 & 2 & -1 & 1 \\ 2 & -1 & -6 & 2 \\ 0 & 1 & 2 & 3 \end{pmatrix}.$$

4. If A is a real symmetric positive-definite matrix show that there exists a real symmetric and positive-definite matrix B such that $B^2 = A$. [*Hint* - diagonalize A and take square roots of the diagonal entries.]

5. If $A \in M_n\mathbb{R}$ is non-singular show that the symmetric matrix AA^t is positive-definite. Let B be a positive-definite square root of A as in problem 4. Show that $T = B^{-1}A$ is orthogonal and hence that $A = BT$ where T is orthogonal and B is positive-definite.

(This is known as a *polar decomposition* of the matrix A.)

6. Show that a hermitian matrix $A \in M_n\mathbb{C}$ is positive-definite if and only if $A = P\bar{P}^t$ for some non-singular $P \in M_n\mathbb{C}$. [*Hint* - there exists a non-singular matrix Q such that $\bar{Q}^tAQ = D$ is diagonal with positive real diagonal entries. Taking a square root of D enables you to write $A = P\bar{P}^t$.]

7. Show that a hermitian $n \times n$ matrix A is negative-definite if and only if the principal minors D_i of A, $i = 1,2,. . ,n$, alternate in sign in the following manner;

$$D_1 < 0, D_2 > 0, D_3 < 0, D_4 > 0, \text{etc.}$$

8. Let $A \in M_n\mathbb{R}$ be positive-definite and symmetric, let $b \in \mathbb{R}^n$ be a column vector of constants, and $x \in \mathbb{R}^n$ a column vector of unknowns. Show that the solution set of the system of linear equations $Ax = b$ coincides with the set of values of x which minimize the function $(1/2)x^tAx - x^tb$.

[*Hint* - write $f(x) = (1/2)x^tAx - x^tb$ and show that $f(y) - f(x) \geq 0$ for all

$y \in \mathbb{R}^n$ and all x satisfying $Ax = b$.]

9. Let $h: V \times V \longrightarrow \mathbb{C}$ be a hermitian form which fails to be non-singular. Let $V_0 = \{v \in V : h(v,w) = 0 \text{ for all } w \in V\}$.

(By problem 1 above $V_0 = 0$ if and only if h is non-singular.)

Show that V has a subspace V_1 such that $V = V_0 \oplus V_1$, a direct sum of vector spaces, where h is non-singular on V_1 while h is identically zero on V_0, i.e. $h(x,y) = 0$ for all x,y in V_0.

6.6 The signature of a quadratic or hermitian form

We now describe a well-known invariant for real quadratic forms and hermitian forms. We use this invariant later to provide a necessary and sufficient condition for two hermitian or real symmetric matrices to be congruent. The basic idea is as follows;

Given a non-singular form on a vector space V we show that V has a direct sum decomposition $V = V^+ \oplus V^-$ where the form is positive-definite on V^+ and negative-definite on V^-. The signature of the form is defined to be the difference in the dimensions of V^+ and V^-.

We state and prove our results in the language of hermitian forms but the reader should be able to adapt them easily to the case of real symmetric bilinear forms.

6.6.1 Lemma

Let λ and μ be two different eigenvalues of a hermitian matrix A. Let v and w be eigenvectors corresponding to λ and μ respectively. Then v and w are orthogonal with respect to the Euclidean inner product on $\mathbb{C}^n$, i.e. $\bar{v}^t w = 0$.

Proof

Since $Aw = \mu w$ we have that $\bar{v}^t A w = \bar{v}^t \mu w = \mu \bar{v}^t w$.

Also $\bar{v}^t A w = \bar{v}^t \bar{A}^t w = (\overline{A v})^t w = (\overline{\lambda v})^t w = \bar{\lambda}\bar{v}^t w = \lambda \bar{v}^t w$ as λ is real by (4.3.6). Since $\lambda \neq \mu$ it follows that $\bar{v}^t w = 0$.

6.6.2 **Proposition**

Let $h: V \times V \longrightarrow \mathbb{C}$ be a non-singular hermitian form. Then there is a direct sum decomposition $V = V^+ \oplus V^-$ such that h restricted to V^+ (resp. V^-) is positive-definite (resp. negative-definite). This decomposition is not unique but if $V = W^+ \oplus W^-$ is another direct sum decomposition with the same property then W^+ has the same dimension as V^+ (and thus W^- has the same dimension as V^-).

Proof

By choosing a basis of V we may, via (2.3.7), identify V with $\mathbb{C}^n$ where n is the dimension of V, and regard h as being given by $h(v,w) = \bar{v}^t A w$ for some hermitian matrix A.

If v is an eigenvector corresponding to some eigenvalue λ of A then $h(v,v) = \lambda \bar{v}^t v$. Now each λ is real by (4.3.6) so that $h(v,v) > 0$ for $\lambda > 0$ while $h(v,v) < 0$ for $\lambda < 0$. By (6.1.4) the matrix A is diagonalizable and so, by (4.6.5), V has a basis consisting of eigenvectors of A. We let V^+ (resp.) V^- denote the direct sum of the eigenspaces corresponding to all of the positive (resp.) negative eigenvalues of A. Then $V = V^+ \oplus V^-$ since zero cannot be an eigenvalue of A. (The matrix A is non-singular since h is non-singular!)

Using (6.6.1), and (2.7.7) if necessary, we can find an orthogonal basis of V^+ consisting entirely of eigenvectors v_i corresponding to positive eigenvalues λ_i of A. If $v \in V^+$ then $v = \Sigma\, \alpha_i v_i$ for some set of scalars α_i in $\mathbb{C}$. It follows that $h(v,v) = \bar{v}^t A v = \Sigma\, \bar{\alpha}_i \alpha_i \lambda_i \bar{v}_i^t v_i$ which is positive for any non-zero vector v. This shows that h is positive-definite on the subspace V^+.

Similarly h is negative-definite on the subspace V^-.

Suppose now that $V = W^+ \oplus W^-$ is another decomposition with the same property. Then $W^+ \cap V^- = 0$ since $h(v,v) > 0$ for $v \in W^+$, $v \neq 0$ while $h(v,v) < 0$ for $v \in V^-$, $v \neq 0$. Consider the quotient space V/W^+. (See (2.4.6) for the definition of quotient space.) The fact that $W^+ \cap V^- = 0$ implies that V^- can be identified with a subspace of V/W^+. (Specifically V^- can be identified with the subspace $\{x + W^+ : x \in V^-\}$.)

Now $\dim (V/W^+) = \dim V - \dim W^+$ by (2.4.7).

$$= \dim W^- \text{ as } V = W^+ \oplus W^-.$$

Hence we have the inequality $\dim V^- \leq \dim W^-$.

Switching the roles of $V^+$and $W^+$and of V^- and W^- in the above argument we obtain the reverse inequality, i.e. $\dim W^- \leq \dim V^-$. Thus $\dim V^- = \dim W^-$ and $\dim V^+ = \dim W^+$. This completes the proof.

6.6.3 **Definition**

The *signature* of a non-singular hermitian form $h: V \times V \longrightarrow \mathbb{C}$ is the integer $\dim V^+ - \dim V^-$ where $V = V^+ \oplus V^-$ is any decomposition of V in the manner of (6.6.2).

The signature is well-defined because of (6.6.2).

Similarly the signature of a non-singular real symmetric bilinear form or real quadratic form is defined.

6.6.4 **Comment**

For a form h which is singular, Proposition (6.6.2) extends to the result that $V = V^+ \oplus V^- \oplus V_0$ where V^+ and V^- are as earlier and $V_0 = \{v \in V : h(v,w) = 0 \text{ for all } w \in V\}$. The form h is zero throughout V_0, i.e. $h(v,w) = 0$ for all $v, w \in V_0$. See also problem 9 of Problems 6C. We may define the signature of h as in (6.6.3).

6.6.5 **Remark on terminology**

Proposition (6.6.2) is often known as "Sylvester's law of inertia". The *inertia* of a form is defined to be the pair of integers (dim V^+,dim V^-), or in the singular case the triple of integers (dim V^+,dim V^-,dim V_0). Sylvester's law says that the inertia is independent of the particular decomposition of V.

6.6.6 **Calculation of the signature**

Let the hermitian or real quadratic form be represented by a matrix A. The signature of the form may be obtained in either of the following two ways;

(i) The signature is r - s where r (resp. s) is the number of positive (resp. negative) eigenvalues of A.

(ii) Reduce A to a real diagonal matrix D by a congruence transformation, i.e. $\bar{P}^t AP = D$ for some non-singular matrix P. Then the signature is r - s where r (resp. s) is the number of positive (resp. negative) entries of D.

There is a straightforward procedure for effecting a congruence transformation of A to diagonal form. The idea is to successively change each off-diagonal entry of A to zero by elementary row and column operations. Because of the symmetry or hermitian symmetry of A we perform a column operation followed at once by the same row operation, (or in the hermitian case the conjugate row operation). If the first operation changes a_{ij} to zero then the second changes a_{ji} to zero. The following example illustrates the technique;

6.6.7 **Example**

Find the signature of the hermitian form represented by the matrix

$$\begin{pmatrix} 1 & i & 2 \\ -i & 3 & 0 \\ 2 & 0 & 4 \end{pmatrix}$$

Subtracting twice column 1 from column 3 followed by subtracting twice row 1 from row 3 yields $\begin{pmatrix} 1 & i & 0 \\ -i & 3 & 2i \\ 0 & -2i & 0 \end{pmatrix}$.

Subtracting i times column 1 from column 2 followed by subtracting $\bar{i}$ times row 1 from row 2 yields $\begin{pmatrix} 1 & 0 & 0 \\ 0 & 4 & 2i \\ 0 & -2i & 0 \end{pmatrix}$.

Subtracting (i/2) times column 2 from column 3 followed by subtracting $(\bar{i}/2)$ times row 2 from row 3 yields $\begin{pmatrix} 1 & 0 & 0 \\ 0 & 4 & 0 \\ 0 & 0 & -1 \end{pmatrix}$.

We now have a diagonal form and the signature is 2 - 1 = 1.

6.6.8 **Signatures and matrices**

We may define the *signature* of a hermitian (or real symmetric) n×n matrix A to be the signature of the hermitian form on $\mathbb{C}^n$ given by $h(v,w) = \bar{v}^t A w$ (or quadratic form on $\mathbb{R}^n$ given by $q(v) = v^t A v$).

6.6.9 **Proposition**

Two non-singular hermitian matrices in $M_n\mathbb{C}$ are congruent if and only if they have the same signature.

Two non-singular real symmetric matrices in $M_n\mathbb{R}$ are congruent if and only if they have the same signature.

Proof

We prove the result in the hermitian case. The real symmetric case is done similarly.

Suppose first that A and B are congruent hermitian matrices. Then $\bar{P}^t AP = B$ for some invertible matrix P. The hermitian form $h : \mathbb{C}^n \times \mathbb{C}^n \longrightarrow \mathbb{C}$, $h(v,w) = \bar{v}^t Aw$, has matrix A with respect to the standard basis $\{e_1, e_2, ., e_n\}$ of $\mathbb{C}^n$ and has matrix B with respect to the basis $\{Pe_1, Pe_2, . . ., Pe_n\}$ of $\mathbb{C}^n$. Hence A and B have the same signature.

Conversely suppose that A and B are non-singular hermitian n×n matrices with the same signature. By (6.4.1) there exists an invertible $P \in M_n\mathbb{C}$ such that $\bar{P}^t AP = D$ where D is real diagonal with entries $\alpha_1, \alpha_2, ., \alpha_r, \beta_1, \beta_2, ., \beta_s$ on the diagonal, each α_i being positive and each β_i being negative. The integer r-s is the signature. Letting Q be the real diagonal matrix with entries $\alpha_1^{-1/2}, . . . , \alpha_r^{-1/2}, (-\beta_1)^{-1/2}, ..., (-\beta_s)^{-1/2}$ we find that $\bar{Q}^t AQ = I_{r,s}$ where $I_{r,s}$ is the diagonal matrix with 1 appearing r times and -1 appearing s times on the diagonal. Thus $(\overline{PQ})^t A(PQ) = I_{r,s}$ so that A is congruent to $I_{r,s}$. Since B has the same signature we see that B is also congruent to $I_{r,s}$. This shows that A and B are congruent.

6.6.10 **Comment**

If A and B in (6.6.9) are singular and they have the same rank then they will be congruent if and only if they have the same signature. In this case each matrix is congruent to $I_{r,s,t}$ which is diagonal with 1 appearing r times, -1 appearing s times, and 0 appearing t times on the diagonal.

6.6.11 **Exercise**

Determine whether or not $\begin{pmatrix} 1 & 2 & 3 \\ 2 & 1 & 0 \\ 3 & 0 & 1 \end{pmatrix}$ and $\begin{pmatrix} 1 & 3 & 2 \\ 3 & 1 & 0 \\ 2 & 0 & 1 \end{pmatrix}$ are congruent.

6.6.12 **Remark**

Unless P is unitary the eigenvalues of a hermitian matrix A will differ from those of $\bar{P}^t AP$, i.e. the eigenvalues are not in general preserved by a congruence transformation. (This is in contrast to the case of similarity transformations which do preserve eigenvalues. See (4.3.2)). Sylvester's law of inertia (6.6.2) ensures that, under a congruence transformation, the signs of the eigenvalues will not change. It is possible to obtain quantitative bounds on the changes in magnitude of the eigenvalues under a congruence transformation but that goes beyond the scope of this book. See [HJ, p224] for more information.

Problems 6D

1. Determine the signature of each of the following quadratic forms;

(i) $q(x,y,z) = 6x^2 + y^2 + z^2 + 2xy + 2yz + 4xz$,

(ii) $q(x,y,z) = 6x^2 - y^2 - z^2 + 2xy + 2yz + 4xz$.

(iii) $q(x,y,z) = 6x^2 - 6y^2 + 2z^2 + 4xy$.

2. If the two real symmetric matrices A and B are congruent show that det A and det B have the same sign.

3. Determine whether or not the following two hermitian matrices are congruent in $M_2\mathbb{C}$;

$$\begin{pmatrix}1 & 0\\0 & -1\end{pmatrix}, \begin{pmatrix}0 & i\\-i & 0\end{pmatrix}$$

4. Let q be the underlying quadratic form of a hermitian form h as in problem 3 of Problems 6B. Show that the signature of q is twice the signature of h.

6.7 Simultaneous reduction of a pair of forms

6.7.1 Proposition

Let A and B be hermitian matrices in $M_n\mathbb{C}$ and assume that A is positive-definite. Then there exists an invertible matrix P and a real diagonal matrix D such that $\bar{P}^tAP = I$ and $\bar{P}^tBP = D$.

(I denotes the identity matrix.)

Proof

A has signature n because it is positive-definite. Hence, by (6.6.9), there exists an invertible matrix Q such that $\bar{Q}^tAQ = I$.

Now apply Schur's theorem (6.1.3) to the hermitian matrix $\bar{Q}^tBQ$ and we obtain a unitary matrix R such that $\bar{R}^t\bar{Q}^tBQR = D$ where D is a real diagonal matrix. The matrix $P = QR$ satisfies the required conditions since $\bar{P}^tBP = D$ and $\bar{P}^tAP = \bar{R}^t\bar{Q}^tAQR = \bar{R}^tIR = I$.

6.7.2 Corollary

Let A, B, P and D be as in (6.7.1). The diagonal entries of D are the eigenvalues of the matrix $A^{-1}B$ and the columns of P are eigenvectors of $A^{-1}B$ corresponding to these eigenvalues.

Proof

Let $p_1, p_2, \ldots, p_n$ denote the columns of P and $d_1, d_2, \ldots, d_n$ the diagonal entries of D. Then $A^{-1}BP = A^{-1}(\bar{P}^t)^{-1}D = PD$ since $\bar{P}^tBP = D$ and $\bar{P}^tAP = I$. Hence $(A^{-1}B)p_i = d_ip_i$ for each i. This proves the corollary.

6.7.3 Note

The above shows that $\mathbb{C}^n$ has a basis consisting entirely of eigenvectors of $A^{-1}B$ and thus, by (4.6.5), $A^{-1}B$ is similar to D. There is no reason however to expect that $A^{-1}B$ is congruent to D.

6.7.4 **Note**

The columns p_i in (6.7.2) are orthonormal with respect to A in the sense that $\bar{p}_i^t A p_j = 1$ if $i = j$, $\bar{p}_i^t A p_j = 0$ if $i \neq j$.

6.7.5 **Exercise**

Let A and B be n×n matrices with A non-singular. Prove that

(i) λ is an eigenvalue of $A^{-1}B$ if and only if $\det(B - \lambda A) = 0$.

(ii) the non-zero vector v is an eigenvector of $A^{-1}B$ if and only if $(B - \lambda A)v = 0$.

6.7.6 **Comment**

In practice to find a matrix P which will perform a simultaneous reduction as in (6.7.1) it is usual to solve the "generalized eigenvalue problem" for $B - \lambda A$ as in (6.7.5). This is easier than directly calculating the eigenvalues and eigenvectors of $A^{-1}B$. The vectors v arising in the solution of (6.7.5) (ii) must be normalized with respect to A, i.e. $\bar{v}^t A v = 1$, in order to be the columns of P. This is because of (6.7.4).

6.7.7 **Example**

Let $A = \begin{pmatrix} 9 & 2 \\ 2 & 1 \end{pmatrix}$ and $B = \begin{pmatrix} 3 & 4 \\ 4 & 2 \end{pmatrix}$. Find P such that $P^t A P = I$ and $P^t B P$ is diagonal.

Notice that A is positive-definite while B is not by (6.5.5).

$$\det(B - \lambda A) = \det\begin{pmatrix} 3-9\lambda & 4-2\lambda \\ 4-2\lambda & 2-\lambda \end{pmatrix} = 5\lambda^2-5\lambda-10 = 5(\lambda-2)(\lambda-1).$$

Solving $(B - \lambda A)v = 0$ when $\lambda = 2$ means solving $\begin{pmatrix} -15 & 0 \\ 0 & 0 \end{pmatrix}\begin{pmatrix} x \\ y \end{pmatrix} = \begin{pmatrix} 0 \\ 0 \end{pmatrix}$. The solution set is $\{ \alpha \begin{pmatrix} 0 \\ 1 \end{pmatrix} : \alpha \in \mathbb{C} \}$. In fact $\begin{pmatrix} 0 \\ 1 \end{pmatrix}$ will do as the first column of P since $(0 \;\; 1)\begin{pmatrix} 9 & 2 \\ 2 & 1 \end{pmatrix}\begin{pmatrix} 0 \\ 1 \end{pmatrix} = 1$.

Solving $(B - \lambda A)v = 0$ for $\lambda = -1$ means solving $\begin{pmatrix} 12 & 6 \\ 6 & 3 \end{pmatrix}\begin{pmatrix} x \\ y \end{pmatrix} = \begin{pmatrix} 0 \\ 0 \end{pmatrix}$.

The solution set is $\{ \alpha\begin{pmatrix}1\\-2\end{pmatrix} : \alpha \in \mathbb{C} \}$. Normalizing with respect to A we take $\begin{pmatrix}1/\sqrt{5}\\-2/\sqrt{5}\end{pmatrix}$ as the second column of P because $(1 \quad -2)\begin{pmatrix}9 & 2\\2 & 1\end{pmatrix}\begin{pmatrix}1\\-2\end{pmatrix} = 5.$

The matrix $P = \begin{pmatrix}0 & 1/\sqrt{5}\\1 & -2/\sqrt{5}\end{pmatrix}$ satisfies the desired properties that $\bar{P}^tAP = I$ and $\bar{P}^tBP = \begin{pmatrix}2 & 0\\0 & -1\end{pmatrix}$. The reader should verify this.

6.7.8 **Comment**

If the method of (6.7.7) is followed except that the vectors v_i are not normalized with respect to A then the matrix with v_i as columns will still be such that $\bar{P}^tAP$ and $\bar{P}^tBP$ are each diagonal but $\bar{P}^tAP$ will no longer equal I.

Even when neither A nor B is positive-definite it may still be possible to find an invertible matrix P such that $\bar{P}^tAP$ and $\bar{P}^tBP$ are each diagonal. This is the case if the polynomial $\det(B - \lambda A)$ has n roots $\lambda_1, \lambda_2, \ldots, \lambda_n$, not necessarily distinct, and if also there exists a linearly independent set of vectors $v_1, v_2, \ldots, v_n$ such that $(B - \lambda_i A)v_i = 0$ for each i. The matrix P with v_i as columns has the required property. See problem 3 of Problems 6E.

Since $\det(B - \lambda A)$ is a polynomial of degree n in λ it will, by the fundamental theorem of algebra (4.2.6), have n roots $\lambda_1, \lambda_2, \ldots, \lambda_n$, not necessarily all different, in $\mathbb{C}$. These roots λ_i are necessarily all real in the situation when A and B are hermitian and A is positive-definite. (See problem 5 of Problems 6E). This, together with (6.7.5), shows that the matrix $A^{-1}B$ has all of its eigenvalues real in the situation when A and B are hermitian with A positive-definite. When A fails to be positive-definite this need not be the case even if A is non-singular.

6.7.9 **Exercise**

Let $A = \begin{pmatrix} 0 & 1 \\ 1 & 0 \end{pmatrix}$ and $B = \begin{pmatrix} 1 & 1 \\ 1 & -1 \end{pmatrix}$. Show that the eigenvalues of $A^{-1}B$ are not real.

6.7.10 **Comment**

The situation of (6.7.1) does indeed have application in physical problems. In the theory of vibrations a system of differential equations of the form $A\ddot{x} + Bx = 0$ can occur, A and B being real symmetric n×n matrices, A positive-definite, $x = (x_i)$ being a column vector of functions x_i, each dependent on time, and $\ddot{x} = (\ddot{x}_i)$ being the column vector of second derivatives.

By (6.7.1) there exists an invertible n×n matrix P such that $P^tAP = I$ and $P^tBP = D$ where D is diagonal with entries $\lambda_1, \lambda_2, .., \lambda_n$ down the diagonal. These λ_i are the solutions of $\det(B - \lambda A) = 0$ as in (6.7.6). After multiplying the system on the left by P^t and substituting $y = P^{-1}x$ we obtain the system $\ddot{y} + Dy = 0$. This latter system consists of n separate equations $\ddot{y}_i + \lambda_i y_i = 0$ which are easy to solve.

Problems 6E

1. Let $A = \begin{pmatrix} 1 & 5 \\ 5 & 26 \end{pmatrix}$ and $B = \begin{pmatrix} 1 & 8 \\ 8 & 56 \end{pmatrix}$. Find P such that $P^tAP = I$ while P^tBP is diagonal.

2. Let $q_1(x,y,z) = 3x^2 + 6y^2 + 2z^2 - 8xy - 6yz + 4xz$,

$q_2(x,y,z) = 3y^2 - 2xy - 2xz$.

Show that q_1 is positive-definite. Find an invertible matrix P such that if

$\begin{pmatrix} x \\ y \\ z \end{pmatrix} = P\begin{pmatrix} u \\ v \\ w \end{pmatrix}$ then $q_1(u,v,w) = u^2 + v^2 + w^2$ and $q_2(u,v,w) = \lambda_1 u^2 + \lambda_2 v^2 + \lambda_3 w^2$ for suitable real numbers λ_1, λ_2, and λ_3.

3. Let $A = \begin{pmatrix} -2 & 1 \\ 1 & -1 \end{pmatrix}$ and $B = \begin{pmatrix} 5 & 2 \\ 2 & -2 \end{pmatrix}$. Show that neither A nor B is positive-definite. Find an invertible 2x2 matrix P such that P^tAP and P^tBP are each diagonal.

4. Solve the following system of differential equations;

$$8\ddot{x}_1 - 3\ddot{x}_2 + 6x_2 = 0$$

$$\ddot{x}_1 - \ddot{x}_2 - 2x_1 - 2x_2 = 0$$

(Rewrite in the form $A\ddot{x} + Bx = 0$ as in (6.7.10).)

5. Let A and B be hermitian nxn matrices and suppose that A is positive-definite. Show that all of the roots of det $(B - \lambda A)$ are real.

[*Hint* - let v be a non-zero vector for which $(B - \lambda A)v = 0$ and show that $\lambda\bar{v}^tAv = \bar{\lambda}\bar{v}^tAv$.]

6.8 Eigenvalues of real symmetric and hermitian matrices

As earlier we will talk in the language of hermitian matrices which includes real symmetric matrices as a special case. We recall two basic facts about hermitian and real symmetric matrices in the following;

6.8.1 Proposition

Let A be a hermitian or real symmetric matrix.

(i) The eigenvalues of A are real.

(ii) Eigenvectors corresponding to distinct eigenvalues of A are orthogonal.

(This orthogonality is with respect to the Euclidean inner product.)

Proof

Statement (i) was proved in (4.3.6) and statement (ii) in (6.6.1).

6.8.2 **Remark**

The symmetry of a hermitian matrix A enables us to change the problem of finding the largest or smallest eigenvalue of A into a problem on maximizing or minimizing real-valued functions. Information on the intermediate eigenvalues is also obtainable by optimization methods. The methods and results which we will describe are often referred to as *variational.*

6.8.3 **Definition**

Let $A \in M_n\mathbb{C}$ be hermitian and let $v \in \mathbb{C}^n$, $v \neq 0$.

The *Rayleigh quotient* for the matrix A at the vector v is the real number $\rho(v)$ given by $\rho(v) = \dfrac{\bar{v}^t Av}{\bar{v}^t v}$.

Note that $\bar{v}^t v \neq 0$ since $v \neq 0$ and that $\rho(v)$ is necessarily real because A is hermitian.

When A is real symmetric we define $\rho(v)$ for non-zero vectors $v \in \mathbb{R}^n$ in the same way, the complex conjugation sign being unnecessary.

6.8.4 **Exercise**

Show that the supremum of the set of values of $\rho(v)$ over all non-zero vectors v is the same as the supremum of $\rho(v)$ over all vectors v of Euclidean norm one. (Compare with exercise (3.3.10).)

6.8.5 **Theorem**

Let $\lambda_1, \lambda_2, \ldots, \lambda_n$ be the eigenvalues of an $n \times n$ hermitian matrix A written in order so that $\lambda_1 \leq \lambda_2 \leq \ldots \leq \lambda_{n-1} \leq \lambda_n$.

Then, for each $v \neq 0$, $\lambda_1 \leq \rho(v) \leq \lambda_n$.

Proof

By (6.1.4) the matrix A is diagonalizable and thus, by (4.6.5), there exists a basis of $\mathbb{C}^n$ (or $\mathbb{R}^n$ in the real symmetric case) which consists entirely of eigenvectors of A. By (ii) of (6.8.1) we can choose this basis to be orthogonal and normalizing each vector then gives an orthonormal basis $v_1, v_2, \ldots, v_n$ with each v_i being an eigenvector corresponding to the eigenvalue λ_i.

Given $v \in \mathbb{C}^n$, $v \neq 0$, we write $v = \sum_{i=1}^n \alpha_i v_i$ for scalars α_i and $\bar{v}^t v = \sum_{i=1}^n \sum_{j=1}^n \bar{\alpha}_i \alpha_j \bar{v}_i^t v_j = \sum_{i=1}^n \bar{\alpha}_i \alpha_i$ by the orthonormality of $v_1, \ldots, v_n$.

Similarly $\bar{v}^t A v = \sum_{i=1} \sum_{j=1} \bar{\alpha}_i \alpha_j \lambda_j \bar{v}_i^t v_j = \sum_{i=1} \bar{\alpha}_i \alpha_i \lambda_i$.

$$\text{Hence } \lambda_n - \rho(v) = \lambda_n - \left(\frac{\sum_{i=1}^n \bar{\alpha}_i \alpha_i \lambda_i}{\sum_{i=1}^n \bar{\alpha}_i \alpha_i} \right) = \frac{\sum_{i=1}^n \bar{\alpha}_i \alpha_i (\lambda_n - \lambda_i)}{\sum_{i=1}^n \bar{\alpha}_i \alpha_i}.$$

Since $\sum_{i=1}^n \bar{\alpha}_i \alpha_i = \|v\|^2 \geq 0$ and $\lambda_n - \lambda_i \geq 0$ for all i, it follows that $\lambda_n - \rho(v) \geq 0$.

$$\text{In a similar fashion } \rho(v) - \lambda_1 = \frac{\sum_{i=1}^n \bar{\alpha}_i \alpha_i (\lambda_i - \lambda_1)}{\sum_{i=1}^n \bar{\alpha}_i \alpha_i}$$

which yields that $\rho(v) - \lambda_1 \geq 0$. This completes the proof.

6.8.6 **Comment**

The above theorem shows that $\rho(v)$ is a lower bound for the largest eigenvalue of A and an upper bound for the smallest eigenvalue of A. Note also that $\rho(v_i) = \lambda_i$ for each eigenvector v_i so that $\lambda_n = \max \{\rho(v) : v \neq 0\}$ and $\lambda_1 = \min \{\rho(v) : v \neq 0\}$, these maximum and minimum values actually being attained for $v = v_n$ and $v = v_1$ respectively.

6.8.7 **Proposition**

If $\| \;\|$ denotes the Euclidean norm on F^n, $F = \mathbb{R}$ or $\mathbb{C}$, then the corresponding operator norm on $M_n F$, which we will denote by $\| \;\|_0$, is given by $\|A\|_0 = \sqrt{\mu}$ where μ is the largest eigenvalue of $\bar{A}^t A$.

Proof

Using the definition of operator norm in (3.3.8) we see that

$$\|A\|_0^2 = \sup_{v\neq 0} \frac{\|Av\|^2}{\|v\|^2} = \sup_{v\neq 0} \frac{\overline{(Av)}^t\,(Av)}{\bar{v}^t v} = \sup_{v\neq 0} \frac{\bar{v}^t\bar{A}^t A v}{\bar{v}^t v} = \sup_{v\neq 0} \rho(v)$$ where $\rho(v)$ is

the Rayleigh quotient at v for the hermitian matrix $\bar{A}^tA$. The result follows by applying comment (6.8.6) to the matrix $\bar{A}^tA$.

6.8.8 **Remark on terminology**

The operator norm $\|\ \|_0$ in (6.8.7) is often referred to as the *spectral norm* on matrices.

6.8.9 **Comment**

Theorem (6.8.5) can be used to estimate the largest and smallest eigenvalues of a hermitian matrix A. If we calculate $\rho(v)$ for a randomly chosen set of vectors v then we will obtain bounds of some kind on the eigenvalues λ_1 and λ_n. They need not be especially good bounds unless we are lucky enough to have chosen v close to an eigenvector for λ_1 or λ_n. In some physical problems, e.g. vibration problems, it is possible from physical considerations to make a good guess at the shape of an eigenvector for λ_1 and sometimes also for λ_n.

6.8.10 **Example**

Let $A = \begin{pmatrix} 5 & 4 & -4 \\ 4 & 5 & 4 \\ -4 & 4 & 5 \end{pmatrix}$. By calculating $\rho(v)$ for vectors v of the form $\begin{pmatrix} 1 \\ t \\ 1 \end{pmatrix}$, where t is real, obtain estimates for the largest and smallest eigenvalues of A.

$$v^tAv = (1\ t\ 1)\begin{pmatrix} 5 & 4 & -4 \\ 4 & 5 & 4 \\ -4 & 4 & 5 \end{pmatrix}\begin{pmatrix} 1 \\ t \\ 1 \end{pmatrix} = (1 + 4t,\ 8 + 5t,\ 1 + 4t)\begin{pmatrix} 1 \\ t \\ 1 \end{pmatrix}$$

$$= 5t^2 + 16t + 2.$$

Hence $\rho(v) = \dfrac{5t^2 + 16t + 2}{t^2 + 2}$ since $v^t v = t^2 + 2$.

To find the maximum and minimum values of this function of t we must equate the derivative $\dfrac{d\rho(v)}{dt}$ to zero.

Now $\dfrac{d\rho(v)}{dt} = \dfrac{(t^2 + 2)(10t + 16) - (5t^2 + 16t + 2)2t}{(t^2 + 2)^2}$ and equating $\dfrac{d\rho(v)}{dt}$ to zero we find that $16t^2 - 16t - 32 = 0$. This gives $t = 2$ or $t = -1$.

When $t = 2$, $\rho(v) = 9$ and when $t = -1$, $\rho(v) = -3$. Hence we may conclude that $\lambda_3 \geq 9$ and $\lambda_1 \leq -3$ if λ_3 and λ_1 are respectively the largest and the smallest eigenvalues of A.

(In fact $\lambda_3 = 9$ and $\lambda_1 = -3$ for this example).

6.8.11 **Example**

Let $A = \begin{pmatrix} 1 & i & i \\ -i & 2 & -1 \\ -i & -1 & 2 \end{pmatrix}$. By considering $\rho(v)$ for vectors v of the form $\begin{pmatrix} t \\ 1 \\ -1 \end{pmatrix}$, where t is real, obtain an estimate for the largest eigenvalue of A.

$$v^t A v = (t, 1, -1)\begin{pmatrix} 1 & i & i \\ -i & 2 & -1 \\ -i & -1 & 2 \end{pmatrix}\begin{pmatrix} t \\ 1 \\ -1 \end{pmatrix} = (t, it + 3, it + 3)\begin{pmatrix} t \\ 1 \\ -1 \end{pmatrix}$$

$$= t^2 + 6$$

Hence $\rho(v) = \dfrac{t^2 + 6}{t^2 + 2}$ since $v^t v = t^2 + 2$.

It is not hard to see that the maximum value over all real t occurs when $t = 0$ and $\rho(v) = 3$ for $t = 0$. Thus we may say that the largest eigenvalue λ_3 of A satisfies $\lambda_3 \geq 3$.

(We should also allow t to take complex values but in fact they will not yield a larger value for $\rho(v)$ in this example. Indeed $\lambda_3 = 3$ is the largest eigenvalue.)

We admit that the reader should be able to directly calculate the eigenvalues in examples (6.8.10) and (6.8.11) by finding the roots of the characteristic polynomial. These examples are meant to be illustrative of the way in which the Rayleigh quotient can be used. Its usefulness is for larger matrices whose eigenvalues are more difficult to determine exactly.

6.8.12 **Rayleigh's principle**

We have seen in (6.8.6) that $\lambda_n = \max\ \{\rho(v)\ ;\ v \neq 0\ \}$ and that $\lambda_1 = \min\ \{\rho(v)\ :\ v \neq 0\}$. There is an extension of this result to the intermediate eigenvalues. It is known as Rayleigh's principle. To understand the principle in general it is helpful to state it first in two special cases.

(i) Assume that we know λ_n , the largest eigenvalue of a hermitian matrix A, and we know an eigenvector v_n corresponding to λ_n. Then the second largest eigenvalue λ_{n-1} of A is given by

$$\lambda_{n-1} = \max\ \{\rho(v)\ :\ v \neq 0 \text{ and } \bar{v}^t v_n = 0\ \}$$

i.e. we take the maximum of $\rho(v)$ over all non-zero vectors v which are orthogonal to v_n.

(ii) Assume that we know λ_1, the smallest eigenvalue of a hermitian matrix A, and we know an eigenvector v_1 corresponding to λ_1. Then the second smallest eigenvalue λ_2 of A is given by

$$\lambda_2 = \min\ \{\rho(v)\ :\ v \neq 0 \text{ and } \bar{v}^t v_1 = 0\ \}$$

i.e. we take the minimum of $\rho(v)$ over all non-zero vectors v which are orthogonal to v_1.

The proof of (i) follows exactly the same pattern as that of (6.8.5) except that the vector v_n is omitted. Similarly for (ii) with the vector v_1 omitted.

The general Rayleigh's principle says that if we know the n-k largest eigenvalues $\lambda_n, \lambda_{n-1}, \ldots, \lambda_{k+1}$ of a hermitian matrix A and we know corresponding eigenvectors $v_n, v_{n-1}, \ldots, v_{k+1}$ then λ_k is given by

$$\lambda_k = \max \{\rho(v) : v \neq 0 \text{ and } \bar{v}^t v_i = 0 \text{ for } i = n, n-1, \ldots, k+1\}$$

i.e. we take the maximum of $\rho(v)$ over all vectors v which are orthogonal to each of $v_n, v_{n-1}, \ldots, v_{k+1}$.

Starting at the other end, if we know the k - 1 smallest eigenvalues $\lambda_1, \lambda_2, \ldots, \lambda_{k-1}$ of A and corresponding eigenvectors $v_1, v_2, \ldots, v_{k-1}$ then the next eigenvalue λ_k is given by

$$\lambda_k = \min \{\rho(v) : v \neq 0 \text{ and } \bar{v}^t v_i = 0 \text{ for } i = 1, 2, \ldots, k-1\}$$

i.e. we take the minimum of $\rho(v)$ over all non-zero vectors v which are orthogonal to each of $v_1, v_2, \ldots, v_{k-1}$.

The proofs are again like that of (6.8.5) but with the appropriate set of vectors v_i being omitted.

6.8.13 **Example**

Let $A = \begin{pmatrix} 1 & i & i \\ -i & 2 & -1 \\ -i & -1 & 2 \end{pmatrix}$ which we examined in Example (6.8.11). Suppose we know that $\lambda_3 = 3$ and that $v_3 = \begin{pmatrix} 0 \\ 1 \\ -1 \end{pmatrix}$. Then we can use Rayleigh's principle to obtain an estimate for λ_2. By Rayleigh's principle $\lambda_2 = \max \rho(v)$, the maximum being over all vectors v orthogonal to $v_3 = \begin{pmatrix} 0 \\ 1 \\ -1 \end{pmatrix}$. The vectors orthogonal to v_3 are of the form $\begin{pmatrix} x \\ y \\ y \end{pmatrix}$ for $x, y \in \mathbb{C}$. Taking x to be real and $y = 1$ it is easy to check that $\rho(v) = 1$. Also taking $x = i$ and $y = 1$ we find $\rho(v) = 7/3 = 2.333$ (to 3 decimal places), and taking $x = i\sqrt{2}$ and $y = 1$ we find that $\rho(v) = 1 + \sqrt{2} = 2.732$ (to 3 decimal places). Thus our estimate is that $\lambda_2 \geq 1 + \sqrt{2}$.

In fact $\lambda_2 = 1 + \sqrt{2}$ and $v_2 = \begin{pmatrix} i\sqrt{2} \\ 1 \\ 1 \end{pmatrix}$ so our estimate could not have been any better.

We can can now carry on and estimate λ_1 by Rayleigh's principle. We must look at vectors orthogonal to both $v_3 = \begin{pmatrix} 0 \\ 1 \\ -1 \end{pmatrix}$ and $v_2 = \begin{pmatrix} i\sqrt{2} \\ 1 \\ 1 \end{pmatrix}$. It is easy to check that a vector orthogonal to both v_3 and v_2 must be a scalar multiple of $v = \begin{pmatrix} -i\sqrt{2} \\ 1 \\ 1 \end{pmatrix}$. (Remember that this orthogonality is with respect to the Euclidean inner product on $\mathbb{C}^3$.) For this vector v an easy calculation yields that $\rho(v) = 1 - \sqrt{2}$ so this is our estimate for λ_1. It is indeed the exact value of λ_1.

6.9 The Courant-Fisher theorem

Rayleigh's principle is not so useful in practice since it requires explicit knowledge of some of the eigenvalues and eigenvectors and we often have no such knowledge. A more useful result is the following theorem which enables us to estimate the eigenvalues λ_k, $k = 1,2,\ldots,n$, of an $n \times n$ hermitian matrix when we have no knowledge whatever of any of the eigenvalues or eigenvectors.

6.9.1 The Courant-Fisher Min-Max theorem

Let $A \in M_nF$ be hermitian (or symmetric in the case when $F = \mathbb{R}$) and let $\lambda_1 \le \lambda_2 \le \ldots \le \lambda_{n-1} \le \lambda_n$ be the eigenvalues of A.

Then for any k, $1 \le k \le n$, $\lambda_k = \min(\max \rho(v))$ where the minimum and maximum are taken as follows;

Choose any n-k vectors $w_1, w_2, \ldots, w_{n-k}$ in F^n and take the maximum of

$\rho(v)$ over all vectors v orthogonal to each of the vectors w_i, $i = 1,2,.,n-k$. Then take the minimum as the set $\{w_1,w_2,\ldots,w_{n-k}\}$ varies over all possible choices of n-k vectors in F^n.

Proof

By (6.1.3) there exists a unitary matrix U such that $UA\bar{U}^t = D$ where D is diagonal with entries $\lambda_1, \lambda_2, \ldots, \lambda_n$ on the diagonal. Hence $A = \bar{U}^t DU$

so that $\rho(v) = \dfrac{\bar{v}^t Av}{\bar{v}^t v} = \dfrac{\bar{v}^t \bar{U}^t DUv}{\bar{v}^t v} = \dfrac{(\overline{Uv})^t D(Uv)}{(\overline{Uv})^t (Uv)}$.

(Note that $(\overline{Uv})^t Uv = \bar{v}^t\bar{U}^t Uv = \bar{v}^t v$ as U is unitary.)

Thus $\rho_A(v) = \rho_D(Uv)$ where ρ_A, resp. ρ_D, denotes the Rayleigh quotient for A, resp. D. Taking the maximum over all vectors v such that v is orthogonal to the chosen set $\{w_1,w_2,\ldots,w_{n-k}\}$ we have that $\max \rho_A(v) = \max \rho_D(Uv)$.

Since U is unitary it is easy to see that v is orthogonal to w_i if and only if Uv is orthogonal to Uw_i. Writing $y = Uv$ we see that $\max \rho_A(v) = \max \rho_D(y)$, the max. on the left being over all v orthogonal to each of $w_1,w_2,\ldots,w_{n-k}$ and the max. on the right being over all y orthogonal to $Uw_1,Uw_2,\ldots,Uw_{n-k}$.

Note that if $y = (y_i)$, a column vector, then $\rho_D(y) = \sum_{i=1}^{n} \lambda_i |y_i|^2$ since D is diagonal with entries $\lambda_1,\lambda_2,\ldots,\lambda_n$. Now suppose we take $\max \rho_D(y)$ over the smaller set of vectors y which, in addition to being orthogonal to each w_i, are unit vectors and have their first k-1 components zero, i.e. $y_1 = y_2 = \ldots = y_{k-1} = 0$

Then $\max \rho_A(v) \geq \max \sum_{i=k}^{n} \lambda_i |y_i|^2$, the max. on the left being as earlier, the max. on the right being as described above.

Hence $\max \rho_A(v) \geq \lambda_k$ since $\lambda_i \geq \lambda_k$ for all $i \geq k$ and $\sum_{i=k}^{n} |y_i^2| = 1$.

This last inequality is valid for all possible choices of the vectors $w_1, w_2, \ldots, w_{n-k}$ and so $\lambda_k \le \min(\max \rho(v))$, the min. and max. being as stated in the theorem.

It remains to show that we have an equality rather than an inequality. By choosing for $\{w_1, w_2, \ldots, w_{n-k}\}$ a set of eigenvectors corresponding to $\lambda_n, \lambda_{n-1}, \ldots, \lambda_{k+1}$ we have, by Rayleigh's principle (6.8.12), that $\lambda_k = \max \rho(v)$ for this particular choice. This completes the proof.

6.9.2 **Corollary**

Let A and B be hermitian n×n matrices. Let $\lambda_k(A)$, $\lambda_k(B)$, and $\lambda_k(A + B)$, $k = 1,2,\ldots,n$, be the eigenvalues of A, B, and A + B respectively, written in ascending order, i.e. $\lambda_1 \le \lambda_2 \le \ldots \le \lambda_n$. Then for each $k = 1,2,\ldots,n$

$$\lambda_k(A) + \lambda_1(B) \le \lambda_k(A + B) \le \lambda_k(A) + \lambda_n(B)$$

Proof

We write $\rho_A(v)$, $\rho_B(v)$, and $\rho_{A+B}(v)$ for the Rayleigh quotient of A, B, and A + B respectively, evaluated at the vector v. It is immediate that $\rho_{A+B}(v) = \rho_A(v) + \rho_B(v)$.

Using (6.9.1), $\lambda_k(A+B) = \min(\max \rho_{A+B}(v))$

$$= \min(\max(\rho_A(v) + \rho_B(v))),$$

where the min. and max. are taken over the sets described in (6.9.1). Hence, using (6.8.5) we see that

$$\lambda_k(A + B) \ge \min(\max(\rho_A(v) + \lambda_1(B)))$$

Now $\min(\max(\rho_A(v) + \lambda_1(B))) = \min(\max \rho_A(v)) + \lambda_1(B)$,

$= \lambda_k(A) + \lambda_1(B)$ after using (6.9.1)

This yields the first half of the inequality ;

$$\lambda_k(A) + \lambda_1(B) \le \lambda_k(A + B)$$

The second half of the inequality is proved in similar fashion.

6.9.3 **Corollary**

Let A and B be hermitian n×n matrices and suppose that $\bar{v}^t Av \le \bar{v}^t Bv$ for all vectors $v \in \mathbb{C}^n$. Let $\lambda_k(A)$ and $\lambda_k(B)$ for $k = 1,2,\ldots,n$ be the eigenvalues of A and B respectively, written in ascending order.

Then $\lambda_k(A) \le \lambda_k(B)$ for each $k = 1,2,\ldots,n$.

Proof

Write $\rho_A(v)$ (resp.$\rho_B(v)$) for the Rayleigh quotient of A (resp.B) at the vector v. The assumption that $\bar{v}^t Av \le \bar{v}^t Bv$ for all v implies that $\rho_A(v) \le \rho_B(v)$ for all $v \ne 0$.

Hence min (max $\rho_A(v)$) $\le$ min (max $\rho_B(v)$), the min. and max. being over the sets described in (6.9.1). Now (6.9.1) yields that $\lambda_k(A) \le \lambda_k(B)$ for each $k = 1,2,\ldots,n$.

(Observe that the condition $\bar{v}^t Av \le \bar{v}^t Bv$ for all $v \in \mathbb{C}^n$ is satisfied when B-A is positive-definite.)

6.9.4 **Remark**

We saw in (6.7.10) that a system of differential equations of the form $A\ddot{x} + Bx = 0$ occurs in the theory of vibrations, A and B being real symmetric matrices with A positive-definite. Let the solutions of $\det(B - \lambda A) = 0$ be λ_i, $i = 1,2,\ldots,n$, where $\lambda_1 \le \lambda_2 \le \ldots \le \lambda_n$. For these vibration problems it turns out that each $\lambda_i \ge 0$ and the possible frequencies of vibration are the values $\sqrt{\lambda_i}$, $i = 1,2,\ldots,n$.

A generalization of the Rayleigh quotient can be used to obtain information about these λ_i. Specifically let $\rho_{A,B}(v) = \dfrac{\bar{v}^t Bv}{\bar{v}^t Av}$ where $v \in \mathbb{R}^n$, and $v \ne 0$.

(Note that $\bar{v}^t Av \ne 0$ for $v \ne 0$ since A is positive-definite.)

It can be shown that a result analogous to (6.8.5) holds for $\rho_{A,B}(v)$ and the "generalized eigenvalues" λ_i as defined above.

i.e. $\lambda_1 \leq \rho_{A,B}(v) \leq \lambda_n$ for all $v \neq 0$. See problems 4 and 5 of Problems 6F.

We should also remark that there is a generalization of the Courant-Fisher theorem which says that $\lambda_k = \min (\max \rho_{A,B}(v))$ for the generalized eigenvalues. See problem 6 of Problems 6F.

Problems 6F

1. Use theorem (6.8.5) to show that a hermitian or real symmetric matrix is positive-definite if and only if each eigenvalue is positive.

2 Let $A = \begin{pmatrix} 0.4 & 0.1 & 0.1 \\ 0.1 & 0.3 & 0.2 \\ 0.1 & 0.2 & 0.3 \end{pmatrix}$. By considering vectors of the form $\begin{pmatrix} x \\ x \\ x \end{pmatrix}$ deduce that the largest eigenvalue of A is 0.6 and by considering vectors of the form $\begin{pmatrix} x \\ y \\ y \end{pmatrix}$ deduce that the second largest eigenvalue is at least 0.3.

3. Let $A = \begin{pmatrix} 6 & 3 & 3\sqrt{2} \\ 3 & 6 & -3\sqrt{2} \\ 3\sqrt{2} & -3\sqrt{2} & 3 \end{pmatrix}$. Let the eigenvalues of A be $\lambda_1,\lambda_2,\lambda_3$ with $\lambda_1 \leq \lambda_2 \leq \lambda_3$. By calculating the Rayleigh quotient for vectors of the form $\begin{pmatrix} 1 \\ -1 \\ x \end{pmatrix}$ deduce that $\lambda_1 \leq -3$. Obtain an estimate for λ_3.

4. Let A and B be real symmetric matrices with A positive-definite. Let

$\rho_{A,B}(v) = \dfrac{\bar{v}^t A v}{\bar{v}^t B v}$ as in (6.9.4).

Let $\lambda_1 \leq \lambda_2 \leq . . . \leq \lambda_n$ be the roots of the polynomial det $(B-\lambda A)$. Show that $\lambda_1 \leq \rho_{A,B}(v) \leq \lambda_n$ for all $v \neq 0$.

5. Let A and B be as in the previous problem. Let C be a real symmetric positive-definite matrix such that $C^2 = A$.

(By problem 4 of Problems 6C such a matrix C exists.)

Show that $\rho_{A,B}(v) = \rho_X(Cv)$ where $X = C^{-1}BC^{-1}$, i.e. the "generalized Rayleigh quotient" $\rho_{A,B}(v)$ equals the ordinary Rayleigh quotient for the matrix $C^{-1}BC^{-1}$ at the vector Cv.

Deduce that $\max_{v \neq 0} \rho_{A,B}(v) = \max_{v \neq 0} \rho_X(v)$, i.e. the largest "generalized eigenvalue" from the equation $\det((B-\lambda A) = 0$ equals the largest eigenvalue of the symmetric matrix $C^{-1}BC^{-1}$.

Obtain a similar result for the smallest "generalized eigenvalue".

6. (Max-Min version of Courant-Fisher theorem)

Let $\lambda_1 \leq \lambda_2 \leq . . . \leq \lambda_n$ be the eigenvalues of the hermitian n×n matrix A. Show that, for each k = 1,2,. . ,n,

$$\lambda_k = \max\ (\min\ \rho(v))$$

where the max. and min. are taken as follows;

Choose any k-1 vectors $w_1, w_2, . . ., w_{k-1}$ and take the minimum of $\rho(v)$ over all vectors v orthogonal to each w_i. Then take the maximum as $\{w_1, w_2, . . ., w_{k-1}\}$ varies through all possible choices of k-1 vectors in $\mathbb{C}^n$.

7. *Hadamard's inequality* says that any $B \in M_n\mathbb{C}$, $B = (b_{ij})$ satisfies

$$\det B \leq \prod_{i=1}^{n} \left(\sum_{j=1}^{n} |b_{ij}|^2\right).$$

(The symbol Π here denotes product)

Obtain this inequality by performing the following series of steps;

(i) If $A = (a_{ij})$ is a positive-definite hermitian n×n matrix show that $\det A \leq \{(1/n)\mathrm{tr}\ A\}^n$.

[*Hint* - use the arithmetic-geometric mean inequality which says that

$(a_1a_2.....a_n)^{1/n} \leq (1/n)(a_1 + a_2 +. . . .+ a_n)$ for any set of positive real numbers $a_1, a_2, . . ., a_n$. Apply it to the eigenvalues of A and use (4.3.5).]

(ii) For A as in (i) show that det A $\leq \prod_{i=1} a_{ii}$.

[*Hint* - if D is diagonal with entries $a_{ii}^{-1/2}$ on the diagonal then

det DAD $= (\prod_{i=1}^{n} a_{ii})^{-1}$ det A so that det A $\leq \prod_{i=1}^{n} a_{ii}$ if and only if det DAD ≤ 1.

Thus it suffices to prove (ii) in the case when each $a_{ii} = 1$.]

(iii) Obtain Hadamard's inequality by putting $A = B\bar{B}^t$ in (ii).

Chapter 7

PERTURBATION THEORY

A small change in some or all of the entries of a matrix is called a *perturbation* of the matrix. The study of the behaviour of matrices under perturbation is known as perturbation theory. Specific questions that we consider in this chapter are ;

(i) when a matrix is perturbed how are its eigenvalues changed ?

(ii) when the matrix of a system of linear equations is perturbed how is the solution set of the system changed ?

We begin the chapter with the theorems of Gershgorin and Taussky which are used for the approximate location in the complex plane of the eigenvalues of a matrix. We consider the problem of whether or not the eigenvalues of a matrix are well-conditioned, i.e. whether or not the changes in the eigenvalues of a perturbed matrix are of the same order as the perturbation. We prove that any complex square matrix can be approximated arbitrarily closely by diagonalizable matrix. Finally we discuss the condition number of a matrix and applications of this to perturbations of linear systems and to inverses of perturbed matrices.

7.1 Gershgorin's theorem and Taussky's theorem

There are certain situations where a precise knowledge of the eigenvalues is not essential. It may be sufficient to know roughly whereabouts in the complex plane the eigenvalues are located. For example the condition that all of the eigenvalues lie in the left-hand half of the complex plane was needed to ensure stability of the solution of a system of differential equations. We encountered this in (5.4.7).

In order to prove a theorem on the approximate location of eigenvalues we first need the following theorem which is of interest in its own right.

7.1.1 **The Dominant Diagonal Theorem**

Let $A = (a_{ij})$ be an nxn matrix with complex entries. Suppose that $|a_{ii}| > \sum_{\substack{j=1 \\ j \neq i}}^{n} |a_{ij}|$ for each i = 1, 2, . . , n. Then A is invertible.

(The condition stated amounts to the fact that the diagonal entry of each row is dominant.)

Proof

We will prove the result by contradiction.

Assuming that A is not invertible there is some non-zero vector $x \in \mathbb{C}^n$ such that Ax = 0. Writing x as a column vector (x_i) the equation Ax = 0 yields a set of n equations ;

$$a_{11}x_1 + a_{12}x_2 + \ldots\ldots + a_{1n}x_n = 0$$

$$a_{21}x_1 + a_{22}x_2 + \ldots\ldots + a_{2n}x_n = 0$$

$$\vdots$$

$$a_{n1}x_1 + a_{n2}x_2 + \ldots\ldots + a_{nn}x_n = 0$$

Choose i such that $|x_i| \geq |x_j|$ for all j = 1, 2, . . , n and examine the i-th equation of the above system.

$$a_{i1}x_1 + a_{i2}x_2 + \ldots\ldots + a_{in}x_n = 0$$

Then $-a_{ii}x_i = \sum_{\substack{j=1 \\ j\neq i}}^{n} a_{ij}x_j$ and properties of the modulus yield that

$$|a_{ii}||x_i| \leq \sum_{\substack{j=1 \\ j\neq i}}^{n} |a_{ij}||x_j|.$$

By our choice of i we have $|x_i| \geq |x_j|$ for all j = 1, 2, . . ., n. Hence for this particular integer i we have $|a_{ii}| \leq \sum_{\substack{j=1 \\ j\neq i}}^{n} |a_{ij}|$. This is contrary to the hypothesis stated in the theorem and hence A must be invertible.

7.1.2 Exercise

Let $A \in M_n\mathbb{R}$ have entries $a_{ij} = 1$ for all $i \neq j$ and $a_{ii} = n$ for each i. Show that A is invertible.

7.1.3 Gershgorin discs

Let $A \in M_n\mathbb{C}$ have entries a_{ij}. For each i = 1, 2, . . . , n we define the *Gershgorin disc* D_i associated to the i-th row of A by

$$D_i = \{ z \in \mathbb{C} : |z - a_{ii}| \leq \sum_{\substack{j=1 \\ j\neq i}}^{n} |a_{ij}| \}.$$

These discs $D_1, D_2, \ldots, D_n$ all lie in the complex plane.

7.1.4 Gershgorin's theorem

Let $A \in M_n\mathbb{C}$. The eigenvalues of A lie in the union of the n Gershgorin discs associated to the rows of A.

Proof

The eigenvalues of A are the roots of the characteristic polynomial $p_A(\lambda) = \det(A - \lambda I)$. If $\lambda_0 \in \mathbb{C}$ is an eigenvalue of A then $A - \lambda_0 I$ is not invertible. Hence, by (7.1.1), λ_0 belongs to at least one of the Gershgorin discs associated to the rows of A.

Warning

This theorem does **not** say that there is one eigenvalue in each disc. See Taussky's theorem below for further information on this.

7.1.5 Example

$$A = \begin{pmatrix} i & 1/2 & 1/2 \\ 1/3 & 2 & 1/2 \\ 1/2 & 0 & -1 \end{pmatrix}$$

The Gershgorin discs for the rows of A are as follows ;

$$D_1 = \{ z \in \mathbb{C} : |z - i| \le 1 \}$$

$$D_2 = \{ z \in \mathbb{C} : |z - 2| \le 5/6 \}$$

$$D_3 = \{ z \in \mathbb{C} : |z + 1| \le 1/2 \}$$

7.1.6 **Remark**

Since A and A^t have the same eigenvalues we may use the columns of A to obtain another set of Gershgorin discs within whose union the eigenvalues of A must lie. This may yield better information.

7.1.7 **Example**

$$A = \begin{pmatrix} 0 & 0.3 & 0.2 & 0.3 & 0.3 \\ -0.1 & 0 & 0.2 & 0.1 & 0.1 \\ 0.2 & 0.1 & 0 & 0.1 & 0.1 \\ 0 & -0.1 & 0.1 & 0 & 0.1 \\ 0.1 & -0.1 & 0.2 & 0.1 & 0 \end{pmatrix}$$

The Gershgorin discs for the rows of A are as follows ;

$$|z| \le 1.1, \quad |z| \le 0.5, \quad |z| \le 0.5, \quad |z| \le 0.3, \quad |z| \le 0.5.$$

The union of these five discs is the single disc $|z| \le 1.1$.

The Gershgorin discs for the columns of A are as follows ;

$$|z| \le 0.4, \quad |z| \le 0.6, \quad |z| \le 0.7, \quad |z| \le 0.6, \quad |z| \le 0.6.$$

The union of these five discs is the single disc $|z| \le 0.7$.

The information from the columns is thus better than that from the rows.

7.1.8 **Continuity of eigenvalues**

An important and useful property of the eigenvalues of a matrix is that they depend continuously on the entries of the matrix. We can state this result in a precise form even though a proof of it is beyond the scope of this book.

7.1.9 **Theorem**

Let A be n×n matrix with entries a_{ij} in $\mathbb{C}$. Let the eigenvalues of A be $\lambda_1, \lambda_2, \ldots, \lambda_n$, not necessarily all different. Then given any real number $\varepsilon > 0$ there exists a real number $\delta > 0$ such that if B is an n×n matrix with entries b_{ij} in $\mathbb{C}$ and $|a_{ij} - b_{ij}| < \delta$ for each i = 1, 2, . . , n then there is labelling $\mu_1, \mu_2, \ldots, \mu_n$ of the eigenvalues of B such that $|\lambda_i - \mu_i| < \varepsilon$ for each i = 1, 2, . . , n.

(Roughly speaking this theorem says that if the entries of A are changed by a sufficiently small amount then the change in the eigenvalues is small.)

The eigenvalues of A are the roots of the characteristic polynomial $p_A(\lambda)$. The coefficients of $p_A(\lambda)$ depend continuously on the entries of A since each coefficient involves sums and differences of products of the entries of A. Theorem (7.1.9) is thus a consequence of a theorem which says that the roots of a polynomial with complex coefficients depend continuously upon the coefficients. This last theorem is usually proved by methods of complex analysis.

7.1.10 **Taussky's theorem**

Let $A = (a_{ij})$ be in $M_n\mathbb{C}$. Suppose that there is an integer k, $1 \leq k \leq n$, such that k of the Gershgorin discs for the rows of A are disjoint from the other n - k discs. Then exactly k eigenvalues of A lie in the union of these k discs.

Proof

Let D be the diagonal matrix with diagonal entries a_{11}, a_{22}, . . .,a_{nn}. Let $M(t) = (1 - t)A + tD$ where $0 \leq t \leq 1$. Then $M(0) = A$ and $M(1) = D$. As t goes from 0 to 1 the matrix M(t) changes continuously from A into D.

The Gershgorin discs for the rows of M(t) are given by

$$|z - a_{ii}| \leq (1 - t) \sum_{\substack{j=1 \\ j \neq i}}^{n} |a_{ij}|$$

for each i = 1, 2, . . , n. The i-th disc for M(t) must lie within the i-th disc for A because $0 \leq 1 - t \leq 1$. Since the eigenvalues of D are its diagonal entries and the eigenvalues vary continuously by (7.1.9), the conclusion of the theorem must hold.

7.1.11 **Example**

$$A = \begin{pmatrix} 2 & 0 & -1 & 0 \\ 0 & -2 & 0 & 1 \\ 0 & 0 & 3 & 1 \\ 1 & 1 & 2 & 5 \end{pmatrix}$$

Show that A has exactly one eigenvalue in the left-hand half-plane.

The Gershgorin discs for the rows of A are as follows ;

$$|z - 2| \leq 1, \ |z + 2| \leq 1, \ |z - 3| \leq 1, \ |z - 5| \leq 4$$

The second disc lies entirely in the left-hand half-plane and is disjoint from the other discs. The result follows at once from Taussky's theorem.

7.1.12 **Special case of Taussky's theorem**

If the Gershgorin discs are mutually disjoint, i.e. if $D_i \cap D_j$ is empty for all $i \neq j$, then there must be exactly one eigenvalue in each disc.

7.1.13 **Exercise**

If $A \in M_n\mathbb{R}$ and the Gershgorin discs for the rows of A are mutually disjoint show that all of the eigenvalues of A must be real.

(*Hint* - since A has real entries the roots of its characteristic polynomial occur in conjugate pairs.)

7.1.14 **Remark**

In the same way as for Gershgorin's theorem we can of course use columns as well as rows to improve our estimates via Taussky's theorem.

The estimates from the Gershgorin and Taussky theorems can be improved by use of diagonal similarity transformations. See problem 3 below.

There are other ways of improving the estimates for the location of eigenvalues, e.g by using rows and columns together in a suitable manner or by using certain kinds of ovals instead of discs. See [HJ, ch. 6] for more information on this.

Problems 7A

1. Show that the converse of the dominant diagonal theorem is false.

(Find an invertible matrix which does not satisfy the dominant diagonal condition.)

2. Show that $A = \begin{pmatrix} 1 & 0 & 1 & 0 \\ 0 & -2 & 0 & 1 \\ 0 & 0 & 3 & 1 \\ 1 & 1 & 2 & 4 \end{pmatrix}$ has exactly one eigenvalue in the left-hand half-plane.

3. Let $A = \begin{pmatrix} 1 & 10^{-5} & 2.10^{-5} \\ 10^{-5} & -1 & -10^{-5} \\ -3.10^{-5} & -10^{-5} & 2 \end{pmatrix}$.

Write down the Gershgorin discs for the rows of A.

Let $S = \begin{pmatrix} 1 & 0 & 0 \\ 0 & 1 & 0 \\ 0 & 0 & \alpha \end{pmatrix}$ where α is a positive real number.

Calculate $S^{-1}AS$ which will have the same eigenvalues as A. Find the largest value of α such that the disc for $S^{-1}AS$ centred at $a_{33} = 2$ does not intersect the other two discs.

Deduce that A has an eigenvalue λ satisfying $|\lambda - 2| \leq 8.10^{-10}$.

4. Let $A = \begin{pmatrix} 5 & \alpha & 4 \\ 0 & 1 & \alpha \\ \alpha & -\alpha & 2 \end{pmatrix}$ where $\alpha \in \mathbb{R}$. Show that if $-1 \leq \alpha \leq 1$ then A has no eigenvalues in the left-hand half-plane.

5. Let B be a 3x3 hermitian matrix for which there exists a unitary matrix U such that $\bar{U}^t BU = \begin{pmatrix} 3.05 & -0.06 & 0.02 \\ -0.06 & -6.91 & 0.07 \\ 0.02 & 0.07 & 8.44 \end{pmatrix}$.

Give the best estimate you can for the eigenvalues of B.

7.2 The condition of eigenvalues

The eigenvalues of a matrix vary continuously with the entries. However they are highly sensitive to change, i.e. a small change in the entries of the matrix can lead to a relatively larger change in the eigenvalues. This does not conflict with the property of being continuous !

7.2.1 Example

$$A = \begin{pmatrix} 0 & 10^4 & 0 & 0 \\ 0 & 0 & 10^4 & 0 \\ 0 & 0 & 0 & 10^4 \\ 0 & 0 & 0 & 0 \end{pmatrix}$$

The characteristic polynomial $p_A(\lambda) = \lambda^4$ so that $\lambda = 0$ is the unique eigenvalue of A occurring with algebraic multiplicity four.

Let $B = \begin{pmatrix} 0 & 10^4 & 0 & 0 \\ 0 & 0 & 10^4 & 0 \\ 0 & 0 & 0 & 10^4 \\ 10^{-4} & 0 & 0 & 0 \end{pmatrix}$. It is easy to verify that $p_B(\lambda) = \lambda^4 - 10^8$

so that the four eigenvalues of B are $\lambda = \pm\, 100, \pm\, 100i$. Thus B has four eigenvalues which are uniformly spaced around the circle of centre 0 and radius 100 in the complex plane. The eigenvalues of B all differ in modulus by a factor of 100 from those of A whereas B differs from A only by a change of $10^{-4} = 0.0001$ in one entry. The change in eigenvalues is of much greater order than the change in the entries.

7.2.2 **Remark**

Example (7.2.1) should suffice to convince the reader that for matrices in general the eigenvalue problem is ill-conditioned. See (3.7) for the notions of "ill-conditioned" and "well-conditioned" numerical problems.

This fact causes difficulty in "real-life" problems. Matrices arising from data obtained in some physical experiment have entries which may be subject to uncertainty due to errors of measurement etc. A small error in the entries may lead to a very much larger error in the eigenvalues of the matrix.

For certain special classes of matrices it is easy to see that the eigenvalue problem is well-conditioned. The classes of diagonal and of triangular matrices are two such examples because the eigenvalues are the diagonal entries. We will see in (8.1.7) and (8.1.8) that the eigenvalue problem is well-conditioned for the classes of real symmetric and complex hermitian matrices.

If the eigenvalues of a matrix are well-conditioned we may ask whether the corresponding eigenvectors are well-conditioned. The separation of the eigenvalues has a great influence on the conditioning of the eigenvectors. If an eigenvalue has algebraic multiplicity one and is isolated then the corresponding eigenvector is well-conditioned. However if the eigenvalue has algebraic multiplicity greater than one or if the eigenvalues are clustered together then the eigenvectors are ill-conditioned. See [GVL, Ch. 7] for further details.

7.2.3 The condition of an individual eigenvalue

In (7.2.1) we saw that the eigenvalue problem is in general ill-conditioned. A matrix can in fact have a mixture of well-conditioned and ill-conditioned eigenvalues. For this reason we need to examine the condition of individual eigenvalues.

Let λ be a simple eigenvalue of the $n \times n$ matrix A, i.e. an eigenvalue of algebraic multiplicity one. Let x be a right eigenvector and $\bar{y}^t$ a left eigenvector of A corresponding to λ. (Thus x and y are column vectors in $\mathbb{C}^n$ and $Ax = \lambda x$, $\bar{y}^t A = \lambda \bar{y}^t$. See also problem 6 of Problems 4A.) Choose x and y to each have euclidean norm one. Let $s(\lambda) = |\bar{y}^t x|$.

7.2.4 Definition

The *condition of the eigenvalue* λ is the real number $1/s(\lambda)$.

(Note that the value of $s(\lambda)$ is uniquely determined because λ is a simple eigenvalue. Note also that $s(\lambda) \leq 1$ by the Cauchy-Schwarz inequality. It can be shown, see problem 5 of Problems 7B, that $s(\lambda) \neq 0$.)

Geometrically $s(\lambda)$ may be interpreted as the cosine of the angle between the vectors x and y.

Now let P be an $n \times n$ matrix with spectral norm one. The spectral norm is the operator norm corresponding to the Euclidean norm on $\mathbb{C}^n$. See (6.8.7). Let us write $A(t) = A + tP$ for the real variable t. Suppose that in the neighbourhood of $t = 0$ there exist differentiable functions $\lambda(t)$ and $x(t)$ such that $A(t)\, x(t) = \lambda(t)\, x(t)$ and $\lambda(0) = \lambda$, $x(0) = x$ where λ and x are as above. Let $\lambda'(t)$ be the derivative of λ with respect to t. Let $x'(t)$ (respectively $A'(t)$) be the vector (respectively matrix) of derivatives of the entries of $x(t)$ (respectively $A(t)$).

7.2.5 **Proposition**

$$\lambda'(0) \le 1/s(\lambda)$$

Proof

Differentiate the equation $A(t)x(t) = \lambda(t)x(t)$ using the product rule and then put $t = 0$ to obtain the equation

$$A'(0)x(0) + A(0)x'(0) = \lambda'(0)x(0) + \lambda(0)x'(0)$$

Thus $Px + Ax'(0) = \lambda'(0)x + \lambda x'(0)$.

Multiplying this equation on the left by $\bar{y}^t$ and using $\bar{y}^t A = \lambda \bar{y}^t$ yields

$$\bar{y}^t Px = \lambda'(0)\ \bar{y}^t x$$

Hence $|\lambda'(0)|\,|\bar{y}^t x| = |y\bar{P}^t x| \le \|\bar{y}^t\|\,\|P\|\,\|x\|$. The right-hand side of this inequality is one by our choice of x, y, and P. The result follows at once.

This last proposition says roughly that a perturbation of order ε in the matrix A leads to a change in the eigenvalue λ of order at most $\varepsilon/s(\lambda)$. If $s(\lambda)$ is close to one then λ is well-conditioned while if $s(\lambda)$ is close to zero then λ is ill-conditioned. When $s(\lambda)$ is close to zero it can be shown that the matrix A is close to a matrix which has λ as a repeated eigenvalue.

7.2.6 **Example**

$$A = \begin{pmatrix} 1 & 1 & 1 \\ 0 & 5 & 1 \\ 0 & 0 & 5.001 \end{pmatrix}$$

The eigenvalues of A are 1, 5, and 5.001.

For $\lambda = 1$ one calculates that $x = \begin{pmatrix} 1 \\ 0 \\ 0 \end{pmatrix}$, $y = \begin{pmatrix} -0.956 \\ 0.239 \\ 0.179 \end{pmatrix}$ and so $s(\lambda) = 0.956$.

For $\lambda = 5$ one calculates that $x = \begin{pmatrix} 0.243 \\ 0.970 \\ 0 \end{pmatrix}$, $y = \begin{pmatrix} 0 \\ 0.001 \\ 0.999 \end{pmatrix}$ and so $s(\lambda) = 0.001$.

$\lambda = 5.001$ one calculates that $x = \begin{pmatrix} 0.243 \\ 0.970 \\ 0.001 \end{pmatrix}$, $y = \begin{pmatrix} 0 \\ 0 \\ 1 \end{pmatrix}$ and so $s(\lambda) = 0.001$.

7.3 **Approximation by diagonalizable matrices**

We will now use the Schur canonical form to show that any matrix in $M_n\mathbb{C}$ can be approximated arbitrarily closely by a matrix which is diagonalizable.

7.3.1 **Lemma**

Let $A = (a_{ij})$ and $B = (b_{ij})$ be in $M_n\mathbb{C}$. If A and B are unitarily equivalent then $\sum_{i=1}^{n}\sum_{j=1}^{n}|a_{ij}|^2 = \sum_{i=1}^{n}\sum_{j=1}^{n}|b_{ij}|^2$, i.e. A and B have the same Euclidean norm.

Proof

$$\bar{A}^t A = \begin{pmatrix} \bar{a}_{11} & \bar{a}_{21} & & \bar{a}_{n1} \\ \bar{a}_{12} & \bar{a}_{22} & & \bar{a}_{n2} \\ & & & \\ \bar{a}_{1n} & \bar{a}_{2n} & & \bar{a}_{nn} \end{pmatrix} \begin{pmatrix} a_{11} & a_{12} & & a_{1n} \\ a_{21} & a_{22} & & a_{2n} \\ & & & \\ a_{n1} & a_{n2} & & a_{nn} \end{pmatrix}$$ and so we see that

trace $\bar{A}^t A = \sum_{i=1}^{n}\sum_{j=1}^{n}|a_{ij}|^2$. Similarly trace $\bar{B}^t B = \sum_{i=1}^{n}\sum_{j=1}^{n}|b_{ij}|^2$.

Since A and B are unitarily similar $B = \bar{P}^t A P$ for some matrix P for which $P^{-1} = \bar{P}^t$. Hence $\bar{B}^t B = \bar{P}^t \bar{A}^t P \bar{P}^t A P = \bar{P}^t \bar{A}^t A P$ and so $\bar{A}^t A$ is similar to $\bar{B}^t B$. This proves the lemma because similar matrices have the same trace .

7.3.2 **Theorem**

Let $A \in M_n\mathbb{C}$ and let $\| \ \|$ be any norm on $M_n\mathbb{C}$. Given any real number $\varepsilon > 0$ there exists a matrix $A_1 \in M_n\mathbb{C}$ which has n distinct eigenvalues and for which $\|A - A_1\| < \varepsilon$.

Proof

We will prove the result using the Euclidean norm on $M_n\mathbb{C}$. Since all norms on $M_n\mathbb{C}$ are equivalent by (3.2.11), the result will follow for any other norm.

Using Schur's theorem (6.1.3) there is a unitary matrix $P \in M_n\mathbb{C}$ and an upper triangular matrix $T \in M_n\mathbb{C}$ such that $\bar{P}^tAP = T$. The eigenvalues of T are its diagonal entries and so we may perturb these entries slightly so as to obtain an upper triangular matrix T_1 with n distinct eigenvalues. We can clearly do this in such a way that $\|T - T_1\| < \varepsilon$.

Let $A_1 = PT_1\bar{P}^t$ which will have the same eigenvalues as T_1. Now since $A = PT\bar{P}^t$ we see that $A - A_1 = P(T - T_1)\bar{P}^t$, i.e. $A - A_1$ and $T - T_1$ are unitarily similar. Thus $\|A - A_1\| = \|T - T_1\|$ by (7.3.1) and the theorem is proved.

7.3.3 **Corollary**

The set of diagonalizable matrices is dense in $M_n\mathbb{C}$, i.e. given any matrix $A \in M_n\mathbb{C}$ there exists a diagonalizable matrix arbitrarily close to A.

Proof

This follows from (7.3.2) and the fact that, by (4.6.6), an $n \times n$ matrix with n distinct eigenvalues is diagonalizable.

7.3.4 Remark

One approach to solving problems involving matrices is firstly to solve the problem for diagonal matrices, secondly for diagonalizable matrices, and finally to approximate a general matrix by a diagonalizable matrix and use some kind of limiting process.

Problems 7B

1. By considering the matrices $\begin{pmatrix} 1 & 0 \\ 0 & 1 \end{pmatrix}$ and $\begin{pmatrix} 1 & 0 \\ \varepsilon & 1 \end{pmatrix}$ show that the geometric multiplicity of an eigenvalue need not be preserved under small perturbations of the entries of the matrix.

2. Let $A = \begin{pmatrix} 1 & 0 \\ 0 & -1 \end{pmatrix}$ and $B = \begin{pmatrix} 1 & \varepsilon \\ \varepsilon & -1 \end{pmatrix}$.

Find the eigenvalues of A and B and deduce that if ε is very small then the eigenvalues of B are very close to those of A.

3. The *spectral radius* $\rho(A)$ of a square matrix A is defined by

$\rho(A) = \max \{ |\lambda| \,;\, \lambda \in \sigma(A) \}$, where $\sigma(A)$ is the spectrum of A.

Let $A = \begin{pmatrix} 1 & \alpha \\ 0 & 1 \end{pmatrix}$ and let $A_\varepsilon = \begin{pmatrix} 1 & \alpha \\ \varepsilon & 1 \end{pmatrix}$ where $\alpha \in \mathbb{R}$, $\varepsilon \in \mathbb{R}$.

Show that $\rho(A) = 1$ and that, for any fixed non-zero value of ε, the value of $\rho(A_\varepsilon)$ tends to infinity as α tends to infinity.

(This example shows the spectral radius can change by an arbitrarily large amount no matter how small the perturbation of the matrix may be.)

4. Let $\| \ \|$ be any norm on $M_n\mathbb{C}$ satisfying property (*) of (3.3.6) and let $S \in M_n\mathbb{C}$ be any invertible matrix. Show that the function $\| \ \|_S$ defined by $\|A\|_S = \|S^{-1}AS\|$ gives a norm on $M_n\mathbb{C}$ which satisfies property (*) of (3.3.6).

Let $\rho(A)$ be the spectral radius of A. Let T be a Schur canonical form for A. Let D be the diagonal matrix with entries $\alpha, \alpha^2, \ldots, \alpha^n$ where α is some real number. Calculate DTD^{-1} and show that, for any $\varepsilon > 0$, we may choose α sufficiently large that $\|DTD^{-1}\|_1 \le \rho(A) + \varepsilon$. (Here $\| \ \|_1$ denotes the maximum absolute column sum norm.)

Deduce that, given any $A \in M_n\mathbb{C}$ and any $\varepsilon > 0$, there exists a norm on $M_n\mathbb{C}$ for which $\rho(A) \le \|A\| \le \rho(A) + \varepsilon$.

5. Let λ be a simple eigenvalue of the $n{\times}n$ matrix A. Let $s(\lambda)$ be the quantity defined in (7.2.4) in the definition of the condition of the eigenvalue λ. Prove that $s(\lambda) \ne 0$.

(*Hint*-there exists a unitary $n{\times}n$ matrix U such that $\bar{U}^tAU$ has λ in the (1,1)-place and zero elsewhere in the first column, the first column of U being x where $Ue_1 = x$ for the standard basis vector e_1. Thus $\bar{U}^tAU$ is of the form $\begin{pmatrix} \lambda & * & * & * & * \\ 0 & & & & \\ \vdots & & \mathbf{B} & & \\ 0 & & & & \end{pmatrix}$ for some (n-1)x(n-1) matrix $\mathbf{B}$. Since λ is a simple eigenvalue of A it cannot be also be an eigenvalue of $\mathbf{B}$. Let $z = \bar{U}^t y$ so that $s(\lambda) = |\bar{y}^t x| = |\bar{z}^t e_1|$. Show that the assumption that $s(\lambda) = 0$ leads to the conclusion that λ is an eigenvalue of $\mathbf{B}$.)

7.4 Condition numbers and their applications

Let $\| \ \|$ be an operator norm on $M_n\mathbb{C}$ and let $A \in M_n\mathbb{C}$ be invertible.

7.4.1 Definition

The *condition number* of A, denoted c(A), is defined by $c(A) = \|A\| \, \|A^{-1}\|$

7.4.2 Remark

The condition number depends on the particular operator norm being used. However the condition numbers of A with respect to two different operator norms are equivalent in a way that is made precise in problem 4 of Problems 7C.

7.4.3 Remark

It is usual to define $c(A) = \infty$ whenever A is not invertible.

7.4.4 Remark

The condition number is useful in studying the behaviour of solutions of systems of linear equations under small perturbations and also for examining how the calculation of the inverse of a matrix is affected by perturbations.

7.4.5 Proposition

Let $Ax = b$ be a system of n linear equations in n unknowns, i.e. A is an nxn matrix, b is a column vector of length n, and x is a column vector of unknowns $x_1, x_2, \ldots, x_n$.

(i) Suppose that A is fixed and b is perturbed to a new vector $b + \delta b$.

If $Ax = b$ and $A(x + \delta x) = b + \delta b$ then $\dfrac{\|\delta x\|}{\|x\|} \le c(A) \dfrac{\|\delta b\|}{\|b\|}$.

(ii) Suppose that b is fixed and A is perturbed to a new matrix $A + \delta A$.

If $Ax = b$ and $(A + \delta A)(x + \delta x) = b$ then $\dfrac{\|\delta x\|}{\|x + \delta x\|} \le c(A) \dfrac{\|\delta A\|}{\|A\|}$.

Proof

(i) $Ax = b$ and $A(x + \delta x) = b + \delta b$ implies that $A\delta x = \delta b$ so that $\delta x = A^{-1}\delta b$. Taking norms yields that $\|\delta x\| \le \|A^{-1}\| \, \|\delta b\|$. Taking norms for the equation $b = Ax$ yields $\|b\| \le \|A\| \, \|x\|$. Multiplying these last two inequalities together yields $\|\delta x\| \, \|b\| \le \|A^{-1}\| \, \|\delta b\| \, \|A\| \, \|x\|$ and hence the desired result that $\dfrac{\|\delta x\|}{\|x\|} \le c(A) \dfrac{\|\delta b\|}{\|b\|}$.

(ii) If $Ax = b$ and $(A + \delta A)(x + \delta x) = b$ then $\delta x = -A^{-1} \, \delta A \, (x + \delta x)$. Taking norms yields that $\|\delta x\| \le \|A^{-1}\| \, \|\delta A\| \, \|x + \delta x\|$. Since $c(A) = \|A\| \, \|A^{-1}\|$ we have the desired result that $\dfrac{\|\delta x\|}{\|x + \delta x\|} \le c(A) \dfrac{\|\delta A\|}{\|A\|}$.

7.4.6 **Lemma**

The condition number $c(A) \ge 1$ for all square matrices A.

Proof

$$1 = \|I\| = \|AA^{-1}\| \le \|A\| \, \|A^{-1}\| = c(A).$$

7.4.7 **Comment**

Proposition (7.4.5) shows that the condition number c(A) gives an upper bound on the error in the solution of the system $Ax = b$ due to an error in the data. If c(A) is close to 1 then the problem is certainly well-conditioned. Since c(A) gives only an upper bound on the size of the error it is possible that the error could be small even if c(A) is large. In general if c(A) is large it means that for some choices of the vector b the system Ax = b will be ill-conditioned.

7.4.8 **Example**

Let $A = \begin{pmatrix} 1 & 10^6 \\ 0 & 1 \end{pmatrix}$ and let $b = \begin{pmatrix} 1 \\ 1 \end{pmatrix}$.

It is easy to check that Ax = b has solution $x = \begin{pmatrix} 1 - 10^6 \\ 1 \end{pmatrix}$.

Changing the vector b to $b + \delta b = \begin{pmatrix} 1 + \delta_1 \\ 1 + \delta_2 \end{pmatrix}$ gives a new solution $x + \delta x$ where

$$\delta x = \begin{pmatrix} \delta_1 - 10^6\delta_2 \\ \delta_2 \end{pmatrix}.$$

Let us use the cartesian norm on $\mathbb{R}^2$ and its corresponding operator norm which, by (3.4.1), is the maximum absolute row sum norm.

Then $\dfrac{\|\delta b\|}{\|b\|} = \max(\,|\delta_1|, |\delta_2|\,)$ since $\|b\| = 1$ and $\dfrac{\|\delta x\|}{\|x\|} = \dfrac{|\,10^6\delta_2 - \delta_1|}{10^6 - 1}$ as we may assume 10^6 is large relative to δ_1 and δ_2.

$$\text{Now } \frac{\|\delta x\|}{\|x\|} = \frac{|\,10^6\delta_2 - \delta_1|}{10^6 - 1} \le \frac{10^6|\delta_2| + |\delta_1|}{10^6 - 1}$$

$$\le \frac{(10^6 + 1)\max(\,|\delta_1|, |\delta_2|)}{10^6 - 1}$$

$$\le \left(\frac{10^6 + 1}{10^6 - 1}\right) \frac{\|\delta b\|}{\|b\|}.$$

Since $\left(\dfrac{10^6 + 1}{10^6 - 1}\right)$ is only fractionally bigger than one it is clear that the error in the solution can be very little larger than the error in the data. However $c(A) = (10^6 + 1)^2$ since $A^{-1} = \begin{pmatrix} 1 & -10^6 \\ 0 & 1 \end{pmatrix}$.

7.4.9 **Exercise**

Prove that $c(AB) \le c(A)\, c(B)$ for any pair of $n \times n$ matrices A and B.

7.4.10 **Definition**

The matrix A is said to be *perfectly conditioned* if $c(A) = 1$.

7.4.11 **Example**

Let $A \in M_n\mathbb{R}$ be an orthogonal matrix so that $\|Av\| = \|v\|$ for all $v \in \mathbb{R}^n$, $\| \ \|$ being the Euclidean norm on $\mathbb{R}^n$. Let us calculate the condition number of A with respect to the spectral norm on $M_n\mathbb{R}$, i.e. the operator norm corresponding to the Euclidean norm on $\mathbb{R}^n$. Then $\|A\| = 1$ by the definition of operator norm and also $\|A^{-1}\| = 1$ since A^{-1} will also be orthogonal. Thus $c(A) = 1$ so that an orthogonal matrix is perfectly conditioned. In fact the only other perfectly conditioned matrices in $M_n\mathbb{R}$ are the scalar multiples of the identity matrix.

7.4.12 **Remark**

If the matrix A fails to be invertible then it is to be expected that the problem of solving the system $Ax = b$ will be ill-conditioned. A small perturbation may change A into an invertible matrix and the nature of the solution set of $Ax = b$ may change completely. The following example illustrates this and gives some motivation for our definition in (7.4.3) that $c(A) = \infty$ when A is not invertible.

7.4.13 **Example**

Let $A = \begin{pmatrix} 1 & 1 \\ 2 & 2 \end{pmatrix}$, $x = \begin{pmatrix} x_1 \\ x_2 \end{pmatrix}$, and $b = \begin{pmatrix} 2 \\ 4 \end{pmatrix}$.

The solution set of $Ax = b$ is infinite, in fact the line $x_1 + x_2 = 2$.

If A is perturbed to $\begin{pmatrix} 1+\varepsilon & 1+\varepsilon \\ 2 & 2 \end{pmatrix}$ then the solution set becomes empty for any $\varepsilon \ne 0$. On the other hand, if A is perturbed to $\begin{pmatrix} 1+\varepsilon & 1 \\ 2 & 2 \end{pmatrix}$ then there is a unique solution, $x_1 = 0$, $x_2 = 2$, for any $\varepsilon \ne 0$.

7.5 Inverses and perturbed matrices

7.5.1 Proposition

Let $A \in M_n F$ be invertible, $F = \mathbb{R}$ or $\mathbb{C}$. If A is perturbed into a matrix $A + P$ where $\|P\| < 1/\|A^{-1}\|$ for some operator norm $\| \ \|$ then $A + P$ is invertible.

Proof

$A + P = A\,(I + A^{-1}P)$ and $\|A^{-1}P\| \le \|A^{-1}\|\ \|P\| < 1$ so that $I + A^{-1}P$ is invertible by the Banach lemma (3.6.1). The result follows since A is also invertible.

7.5.2 Remark

This result shows that if a non-singular matrix is perturbed by a small enough amount it will remain non-singular. This is useful in practical situations where the entries of A may be subject to small errors of measurement.

7.5.3 Proposition

Let $A \in M_n F$ be non-singular and suppose that A is perturbed to $A + \delta A$ where $\|A^{-1}\|\ \|\delta A\| < 1$. Then the matrix $A + \delta A$ is non-singular and

$$\frac{\|A^{-1} - (A + \delta A)^{-1}\|}{\|A^{-1}\|} \le \left(\frac{1}{1 - k}\right) c(A)\,\frac{\|\delta A\|}{\|A\|}$$

where $k = c(A)\,\dfrac{\|\delta A\|}{\|A\|}$.

Proof

Letting $P = \delta A$ in (7.5.1) we see that $A + \delta A$ is non-singular.

$$(A + \delta A)^{-1} = (A\,(I + A^{-1}\delta A))^{-1} = (I + A^{-1}\delta A)^{-1}A^{-1} = \left(\sum_{n=0}^{\infty} (-1)^n (A^{-1}\delta A)^n\right) A^{-1}$$

on using (3.6.2). (Note that $I + A^{-1}\delta A$ is invertible by (3.6.1) since $\|A^{-1}\delta A\| \le \|A^{-1}\|\ \|\delta A\| < 1$.)

Hence $A^{-1} - (A + \delta A)^{-1} = -(\sum_{n=1}^{\infty} (-1)^n(A^{-1}\delta A)^n)\, A^{-1}$ which yields that

$$\| A^{-1} - (A + \delta A)^{-1}\| \le \| \sum_{n=1}^{\infty} (-1)^n(A^{-1}\delta A)^n \| \, \|A^{-1}\|$$

$$\le (\sum_{n=1}^{\infty} \| (A^{-1}\delta A)\|^n)\, \|A^{-1}\|.$$

On summing the geometric series $\sum_{n=1}^{\infty} \| (A^{-1}\delta A)\|^n$ where $\| (A^{-1}\delta A)\| < 1$ we obtain $\| A^{-1} - (A + \delta A)^{-1}\| \le \left(\dfrac{\|A^{-1}\delta A\|}{1 - \|A^{-1}\delta A\|}\right) \|A^{-1}\|$.

Hence $$\frac{\| A^{-1} - (A + \delta A)^{-1}\|}{\|A^{-1}\|} \le \frac{\|A^{-1}\delta A\|}{1 - \|A^{-1}\delta A\|}.$$

Now $k = c(A) \dfrac{\|\delta A\|}{\|A\|} = \|A^{-1}\| \, \|\delta A\| \ge \|A^{-1}\delta A\|$ so that $1 - k \le 1 - \|A^{-1}\delta A\|$ and

$\dfrac{1}{1 - \|A^{-1}\delta A\|} \le \dfrac{1}{1 - k}$. This now yields the desired result.

$$\frac{\| A^{-1} - (A + \delta A)^{-1}\|}{\|A^{-1}\|} \le \left(\frac{1}{1 - k}\right) c(A) \frac{\|\delta A\|}{\|A\|}$$

This proposition shows that for a small perturbation in A the relative change in the inverse of A is of the same order as the relative change in A, provided that this relative change in A is small and c(A) is fairly small.

Problems 7C

1. Let $A = \begin{pmatrix} 1.1 & 2.1 & 3.1 \\ 1 & -1 & 2 \\ 0.2 & 3.3 & 1.4 \end{pmatrix}$.

Calculate the condition number c(A) with respect to each of the following norms ;

(i) the maximum absolute row sum norm,

(ii) the maximum absolute column sum norm.

2. Let $A_n = \begin{pmatrix} 1/n & 1 \\ 1 & 1/n \end{pmatrix}$. Calculate c(A) with respect to the absolute row sum norm and verify that $c(A_n)$ tends to one as n tends to infinity.

3. Let $A = \begin{pmatrix} \alpha & 0 & 1 \\ 0 & \alpha & 0 \\ 1 & 0 & \alpha \end{pmatrix}$ where α is a positive real number.

Calculate c(A) with respect to the maximum absolute row sum norm and verify that c(A) tends to one as α tends to infinity.

4. Let c_1 and c_2 denote the condition numbers with respect to two different operator norms on $M_n F$. Show that there are positive constants k_1 and k_2 such that the following inequalities hold for all $A \in M_n F$;

$$c_1(A) \leq k_1 \; c_2(A)$$
$$c_2(A) \leq k_2 \; c_1(A)$$

(Thus c_1 and c_2 are equivalent in a similar manner to the equivalence of norms as defined in (3.2.10).)

5. Let $A \in M_n \mathbb{C}$ be diagonalizable, i.e. $S^{-1}AS = D$ for some invertible matrix S and some diagonal matrix D. Let D have diagonal entries $\lambda_1, \lambda_2, \ldots, \lambda_n$ which are necessarily the eigenvalues of A. Let A be perturbed to a new matrix $A + \delta A$. Show that if μ is an eigenvalue of $A + \delta A$ then, for some i, $|\mu - \lambda_i| \leq c(S) \; \|\delta A\|$ where the norm being used is the maximum absolute row sum norm.

[*Hint* - apply Gershgorin's theorem to the matrix $S^{-1}(A + \delta A)S = D + S^{-1}\delta AS$.]

6. Let $Ax = b$ be a system of n linear equations in n unknowns. Suppose that A is perturbed to $A + \delta A$ and b to $b + \delta b$. Let $x + \delta x$ be the solution of the perturbed system, i.e. $(A + \delta A)(x + \delta x) = b + \delta b$. Let $M = (1 - \|A^{-1}\delta A\|)^{-1}$.

Show that the following inequality holds ;

$$\frac{\|\delta x\|}{\|x\|} \leq M\, c(A) \left(\frac{\|\delta b\|}{\|b\|} + \frac{\|\delta A\|}{\|A\|} \right)$$

7. Let $p(x) = a_0 + a_1 x + a_2 x^2 + \ldots\ldots + a_{n-1} x^{n-1} + x^n$ be a monic polynomial with coefficients $a_i \in \mathbb{C}$. Let C be the companion matrix of p(x) as defined in problem 7 of Problems 4A.

Write $C = R + S$ where $R = \begin{pmatrix} 0 & 0 & & 0 \\ & & & \\ 0 & 0 & & 0 \\ -a_0 & -a_1 & & -a_{n-1} \end{pmatrix}$ and $S = \begin{pmatrix} 0 & 1 & 0 & & 0 \\ 0 & 0 & 1 & & \\ & & & & \\ 0 & 0 & & 0 & 1 \\ 0 & 0 & & 0 & 0 \end{pmatrix}$.

Show that $\bar{R}^t S = 0 = \bar{S}^t R$, that $\|R\bar{R}^t\| = \sum_{i=0}^{n-1} |a_i|^2$, and that $\|\bar{S}^t S\| = 1$ where $\|\ \|$ is the spectral norm, i.e. the operator norm arising from the Euclidean norm.

Deduce that $\|C\|^2 = \|\bar{C}^t C\| \leq \|\bar{R}^t R\| + \|\bar{S}^t S\|$.

Hence show that if x_0 is any root of the polynomial p(x) then

$$|x_0| \leq (1 + |a_0|^2 + |a_1|^2 + \ldots + |a_{n-1}|^2)^{1/2}$$

8. Let λ be a simple eigenvalue of a hermitian matrix. Show that the condition of λ equals one. (The condition of an eigenvalue is defined in (7.2.4).)

9. Let $A = \begin{pmatrix} 1 & t & t^2 \\ t & 1 & t \\ t^2 & t & 1 \end{pmatrix}$ where $t \in \mathbb{R}$, $0 \leq t < 1$.

Show that the condition number $c(A) = \dfrac{(1 + t)(1 + 2t)}{1 - t}$, using the absolute column sum norm.

10. Let $A = \begin{pmatrix} 1 & a & a^2 \\ 1 & b & b^2 \\ 1 & c & c^2 \end{pmatrix}$ where $a, b, c \in \mathbb{R}$ with $0 < a < b < c < 1/2$.

Show that the condition number $c(A) \geq 8\ (1 + c + c^2)$, using the maximum absolute column sum norm.

Chapter 8

FURTHER TOPICS

In this final chapter we deal with a variety of topics, not necessarily related to each other. We begin by discussing normal matrices and their properties. The notion of a normal matrix includes as special cases real symmetric matrices, hermitian matrices, orthogonal matrices, and unitary matrices. We next describe the QR-factorization of a matrix and the important QR-algorithm which is the most widely used method for the numerical computation of the eigenvalues of a matrix. In the next section we introduce the idea of the singular values of a matrix and derive the singular value decomposition. We show why this decomposition is important in particular in relation to the the numerical calculation of the rank of a matrix. The next two sections deal with generalized inverses of matrices and least squares problems. In the last section of the chapter we consider iterative methods for solving a system of simultaneous linear equations $Ax = b$. In particular we describe the Gauss-Seidel method. These iterative methods are to be preferred to Gaussian elimination when the matrix A is large and sparse, i.e. when A has a large number of its entries equal to zero.

8.1 Normal matrices and their eigensystems

8.1.1 Definition

The matrix $A \in M_n\mathbb{C}$ is *normal* if $A\bar{A}^t = \bar{A}^tA$, i.e. a normal matrix is one which commutes with its conjugate transpose. If $A \in M_n\mathbb{R}$ then A is normal if it commutes with its transpose.

8.1.2 Exercise

Show that each of the following is normal;

(i) a unitary matrix (i.e. $A^{-1} = \bar{A}^t$),

(ii) a real orthogonal matrix (i.e. $A \in M_n\mathbb{R}$ and $A^{-1} = A^t$),

(iii) a hermitian matrix (i.e. $A = \bar{A}^t$),

(iv) a real symmetric matrix (i.e. $A \in M_n\mathbb{R}$ and $A = A^t$).

8.1.3 Theorem

The matrix $A \in M_n\mathbb{C}$ is normal if and only if A is unitarily similar to a diagonal matrix.

Proof

Suppose that $\bar{P}^tAP = D$ where P is unitary and D is diagonal. Then $A = PD\bar{P}^t$ so that $A\bar{A}^t = PD\bar{P}^tP\bar{D}^t\bar{P}^t = PD\bar{D}^t\bar{P}^t$.

Also $\bar{A}^tA = P\bar{D}^t\bar{P}^tPD\bar{P}^t = P\bar{D}^tD\bar{P}$. Now $D\bar{D}^t = \bar{D}^tD$ since D and $\bar{D}^t$ are each diagonal and thus A is normal.

Conversely suppose that A is normal. By Schur's theorem (6.1.3) there exists a unitary matrix P and an upper triangular matrix T such that $\bar{P}^tAP = T$. Let T have entries t_{ij}. We know that $t_{ij} = 0$ for all $i > j$ as T is upper triangular. We must show that also $t_{ij} = 0$ for all $i < j$.

Now $T\bar{T}^t = \bar{P}^tAP\bar{P}^t\bar{A}^tP = \bar{P}^tA\bar{A}^tP = \bar{P}^t\bar{A}^tAP$ as A is normal

$$= \bar{P}^t\bar{A}^tP\bar{P}^tAP = \bar{T}^tT.$$

We multiply out $T\bar{T}^t$ and $\bar{T}^tT$ and equate corresponding diagonal entries. Equating the (1,1)-entries yields that $\sum_{j=1}^{n} |t_{1j}|^2 = |t_{11}|^2$. Hence

$\sum_{j=2}^{n}|t_{1j}|^2 = 0$ from which it follows that $t_{1j} = 0$ for each $j = 2,3,\ldots,n$.

Equating the (2,2)-entries yields $\sum_{j=2}^{n}|t_{2j}|^2 = |t_{12}|^2 + |t_{22}|^2$

But we have shown that $t_{12} = 0$ so that $\sum_{j=3}^{n}|t_{2j}|^2 = 0$ from which it follows that $t_{2j} = 0$ for each $j = 3,4,\ldots,n$.

Proceeding in this fashion with the other diagonal entries we find that $t_{ij} = 0$ for all $i < j$, i.e. T is a diagonal matrix.

8.1.4 **Note**

The argument in the above proof shows that a triangular matrix is normal if and only if it is diagonal.

8.1.5 **Remark on terminology**

Theorem (8.1.3) is often called *the spectral theorem for normal matrices.*

8.1.6 **Exercise**

Show that $A \in M_n\mathbb{C}$ is normal if and only if $\mathbb{C}^n$ has an orthonormal basis consisting entirely of eigenvectors of A.

[*Hint* - examine the proof of (4.6.5) and adapt it to this situation. See also (ii) of (6.8.1).]

Note that in particular this last exercise shows that the eigenvectors corresponding to distinct eigenvalues of a normal matrix are orthogonal.

8.1.7 **Theorem**

Let $A \in M_n\mathbb{C}$ be normal with eigenvalues $\lambda_1,\lambda_2,\ldots,\lambda_n$. Let μ be an eigenvalue of the perturbed matrix $A + \delta A$. Then, for some i, $\|\mu - \lambda_i\| \leq \|\delta A\|$ where $\|\quad\|$ is the spectral norm, (i.e. the operator norm with respect to the euclidean norm on $\mathbb{C}^n$.)

Proof

By (8.1.3) there is a unitary matrix P such that $\bar{P}^t AP = D$ where D is diagonal with entries $\lambda_1, \lambda_2, \ldots, \lambda_n$.

If $(A + \delta A)v = \mu v$ for $\mu \in \mathbb{C}$ and $v \neq 0$ then $(\delta A)v = (\mu I - A)v = (\mu I - PD\bar{P}^t)v = P(\mu I - D)\bar{P}^t v$. Now if $\mu I - D$ is not invertible then $\mu = \lambda_i$ for some i so our theorem is trivially true.

If $\mu I - D$ is invertible then $\bar{P}^t v = (\mu I - D)^{-1}\bar{P}^t(\delta A)v$. Taking euclidean norms yields that $\|\bar{P}^t v\| = \|(\mu I - D)^{-1}\bar{P}^t(\delta A)v\|$

$$\leq \|(\mu I - D)^{-1}\| \, \|\bar{P}^t\| \, \|\delta A\| \, \|v\|$$

after using the properties of the operator norm.

Now $\|\bar{P}^t v\| = \|v\|$ and $\|P\| = 1$ since P is unitary and thus our inequality becomes $1 \leq \|(\mu I - D)^{-1}\| \, \|\delta A\|$. The spectral norm of a diagonal matrix is the maximum absolute value of its diagonal entries and so we have that $\|(\mu I - D)^{-1}\| = \max_i (|\mu - \lambda_i|^{-1}) = (\min_i |\mu - \lambda_i|)^{-1}$. Thus $\min_i |\mu - \lambda_i| \leq \|\delta A\|$ and this completes the proof.

8.1.8 **Comment**

The above shows that the eigenvalues of a normal matrix are well-conditioned. In particular this applies to real symmetric matrices and hermitian matrices. See also problem 3 of Problems 8A.

Problems 8A

1. Show that $\begin{pmatrix} a & -a \\ a & a \end{pmatrix}$ is normal for any $a \in \mathbb{R}$.

2. Show that the matrix $A = (a_{ij})$ is normal if and only if $\sum_{i,j} |a_{ij}|^2 = \sum_{i=1}^{n} |\lambda_i|^2$ where $\lambda_1, \lambda_2, \ldots, \lambda_n$ are the eigenvalues of A, the λ_i not necessarily being distinct.

3. Let A be a normal matrix. Show that v is a right eigenvector of A for the eigenvalue λ if and only if v is a left eigenvector of A for the value λ.

(i.e. show that $Av = \lambda v$ if and only if $\bar{v}^t A = \lambda \bar{v}^t$)

Deduce that the condition $s(\lambda) = 1$ for any simple eigenvalue of a normal matrix A.

4. Let $A = B^{-1}\bar{B}^t$ for some $B \in M_n\mathbb{C}$. Show that A is unitary if and only if B is normal.

Let B be expressible in the form $B = HNH$ where H is hermitian and N is normal. Show that $A = B^{-1}\bar{B}^t$ is similar to a unitary matrix.

5. Let $A \in M_n\mathbb{C}$. Let $H = (1/2)(A + \bar{A}^t)$ and $K = (1/2)(A - \bar{A}^t)$.

Show that A is normal if and only if $HK = KH$.

8.2 The QR-factorization and the QR-algorithm

8.2.1 Eigenvalues and triangularization

We mentioned in chapter 4 that the QR-algorithm is the most commonly used numerical method for determining the eigenvalues of a matrix. In this section we give a brief outline of the ideas behind this algorithm.

Note first that if a matrix A is similar to a diagonal matrix T then the eigenvalues of A are the diagonal entries of T. Unitary triangularization (i.e. as in the reduction to the Schur canonical form) is good for numerical computation because;

(i) $\bar{P}^t$ is much easier to compute than P^{-1}.

(ii) Similarity via a unitary matrix preserves distances between matrices when measured using the spectral norm as in (6.8.8). See the exercise below.

8.2.2 Exercise

Let A and B be in $M_n\mathbb{C}$ and let $P \in M_n\mathbb{C}$ be unitary. Let $\| \; \|$ be the spectral norm on $M_n\mathbb{C}$. Prove that $\| A - B \| = \| \bar{P}^t AP - \bar{P}^t BP \|$.

[*Hint* - use property (*) of the spectral norm and the fact that $\| P \| = 1$.]

8.2.3 Proposition (The QR-factorization)

Let $A \in M_n\mathbb{C}$. Then there exists a unitary matrix $Q \in M_n\mathbb{C}$ and an upper triangular matrix R such that $A = QR$.

Proof

First suppose A is non-singular and let $v_1, v_2, \ldots, v_n$ be the columns of A. Then $\{v_1, v_2, \ldots, v_n\}$ is a basis of $\mathbb{C}^n$. Transform this basis into an orthonormal basis $\{w_1, w_2, \ldots, w_n\}$ by the Gram-Schmidt process (2.7.7). The procedure in the Gram-Schmidt process is such that w_1 is a scalar multiple of v_1, w_2 is a linear combination of v_1 and v_2, w_3 is a linear combination of v_1, v_2, and v_3, etc. Solving for each v_i in terms of the w_i we see that

$$v_1 = r_{11}w_1$$

$$v_2 = r_{12}w_1 + r_{22}w_2$$

$$v_3 = r_{13}w_1 + r_{23}w_2 + r_{33}w_3$$

$$.$$

$$.$$

$$v_n = r_{1n}w_1 + r_{2n}w_2 + \ldots + r_{nn}w_n$$

for suitable scalars $r_{ij} \in \mathbb{C}$.

We let R be the upper triangular matrix with entries r_{ij} and Q the unitary matrix with $w_1, w_2, \ldots, w_n$ as its columns. It is easy to see that $A = QR$.

If A is singular then the columns $v_1, v_2, \ldots, v_n$ of A will be linearly dependent. We can still follow the same mechanical procedure of the Gram-Schmidt process but some of the w_i may turn out to be zero. The set of all non-zero w_i arising will form a linearly independent set and this set can be extended to an orthonormal basis of $\mathbb{C}^n$. Let Q have these basis vectors as its columns and construct R exactly as before. The new vectors adjoined to the set of non-zero w_i are independent of the v_i and so correspond to zero rows in the matrix R. Thus we obtain a factorization $A = QR$ as required.

8.2.4 **Example**

Find a QR-factorization for the matrix $A = \begin{pmatrix} 3 & 2 \\ 4 & 5 \end{pmatrix}$.

Using the above notation $v_1 = \begin{pmatrix} 3 \\ 4 \end{pmatrix}$, $v_2 = \begin{pmatrix} 2 \\ 5 \end{pmatrix}$. Performing the Gram-Schmidt process we find that $w_1 = \begin{pmatrix} 3/5 \\ 4/5 \end{pmatrix}$, $w_2 = \begin{pmatrix} -4/5 \\ 3/5 \end{pmatrix}$.

Hence we may take $Q = \begin{pmatrix} 3/5 & -4/5 \\ 4/5 & 3/5 \end{pmatrix}$.

To find R we must write the vectors v_1 and v_2 in terms of w_1 and w_2. Clearly $v_1 = 5w_1$. Writing $v_2 = \alpha w_1 + \beta w_2$ we have that

$$2 = (3/5)\alpha - (4/5)\beta$$

$$5 = (4/5)\alpha + (3/5)\beta$$

This gives $\alpha = 26/5$ and $\beta = 7/5$. Hence $R = \begin{pmatrix} 5 & 26/5 \\ 0 & 7/5 \end{pmatrix}$ and $A = QR$ is the desired factorization.

8.2.5 **Comment**

For numerical computation the classical Gram-Schmidt algorithm is not so good, the w_i obtained sometimes being far from orthogonal (due to rounding errors etc.) There is a re-arranged version known as the *modified Gram-Schmidt algorithm* which is much better for numerical computation. In fact for a fast effective way to obtain a QR-factorization it is usual to first reduce the matrix to Hessenberg or tridiagonal form, (see (1.5)), via transformations known as *Householder transformations*. See problems 2 and 3 of Problems 8B.

8.2.6 **The QR-algorithm**

Starting with $A \in M_n\mathbb{C}$ we construct recursively a sequence of n×n matrices $\{A_k\}$, $k = 0,1,2.........$,as follows;

Define $A_0 = A$.

For each k we write $A_k = Q_k R_k$ where Q_k is unitary and R_k is upper triangular. Then define $A_{k+1} = R_k Q_k$. (Note that (8.2.3) ensures that a factorization $A_k = Q_k R_k$ is always possible!)

It is clear that each A_k is unitarily similar to A. Specifically $A_k = P^t AP$ where $P = Q_0 Q_1 Q_2 \ldots\ldots Q_k$.

It is not obvious but it can be proved that for a very large class of matrices in $M_n\mathbb{C}$ the matrix A_k tends to an upper triangular matrix as k tends to infinity, (i.e. A_k tends to a Schur canonical form for A as k tends to infinity.)

In particular this is true for any matrix A whose eigenvalues have distinct absolute values. The eigenvalues of A are immediately obtainable as they will equal the diagonal entries of any Schur canonical form for A.

If $A \in M_n\mathbb{R}$ has any complex eigenvalues they will occur in conjugate pairs. We have seen in problem 3 of Problems 6A that any $A \in M_n\mathbb{R}$ has a *real Schur canonical form*, i.e. a block upper triangular matrix whose diagonal blocks are either 1x1 blocks or 2x2 blocks having a pair of complex conjugate eigenvalues. The sequence $\{A_k\}$ of (8.2.6) will tend towards a real Schur canonical form for A as k tends to infinity. The eigenvalues of A are now easily obtainable from those of the real Schur form.

We have only presented a brief outline of the QR-algorithm. The reader who is interested in the technical details of all this should consult [GVL,chapter 7].

We should remark that the QR-algorithm is the most widely used method for numerical calculation of the eigenvalues of a matrix which has no symmetry properties. For a real symmetric matrix the QR-algorithm can be specialized into an algorithm known as the *symmetric QR-algorithm*. There are also several other algorithms which are used for the numerical computation of the eigenvalues of real symmetric matrices. These methods make use of the special properties of symmetric matrices. See [GVL,chapter 8] for details.

Problems 8B

1. Find a QR-decomposition for the matrix $\begin{pmatrix} 1 & -1 & 1 \\ 1 & 0 & 1 \\ -1 & 1 & 1 \end{pmatrix}$.

2. Let $v \in \mathbb{C}^n$ be a non-zero column vector so that $v\bar{v}^t$ is an nxn matrix. Let I be the identity nxn matrix. Let $H = I - (2/\bar{v}^t v)v\bar{v}^t$ which is an nxn matrix.

Show that (i) $Hv = -v$,

(ii) $Hw = w$ for any vector w which is orthogonal to v, (i.e. for which $\bar{w}^t v = 0$.)

(iii) H is a unitary matrix.

(H is called a *Householder matrix* or *Householder transformation*.)

3. Let $A = \begin{pmatrix} 1 & 0 & 1 \\ 0 & 1 & 1 \\ 1 & 1 & 2 \end{pmatrix}$. Let H be the 2x2 Householder matrix given by the vector $v = \begin{pmatrix} 1 \\ 1 \end{pmatrix}$. Let $U = \begin{pmatrix} 1 & 0 & 0 \\ 0 & & \\ 0 & H & \end{pmatrix}$.

Show that $U^t A U$ is tridiagonal.

8.3 The singular value decomposition

8.3.1 Rectangular matrices

For much of the time in this book we have been dealing with square matrices. We will now talk about the singular value decomposition for mxn matrices. Square matrices are of course included as the special case when $m = n$.

Let A be an mxn matrix with complex entries. We may view A as a linear operator $\mathbb{C}^n \longrightarrow \mathbb{C}^m$. This linear operator has an operator norm defined exactly as in (3.3.8). Note that property (*) of (3.3.6) holds for operator norms on rectangular matrices, i.e. $\|AB\| \le \|A\| \, \|B\|$ whenever A is an $m{\times}n$ matrix and B is an $n{\times}r$ matrix. (The proof of this property goes exactly like that in (3.3.11) for square matrices.)

8.3.2 **Lemma**

Let A be an $m{\times}n$ matrix with complex entries. Then all of the eigenvalues of the $n{\times}n$ hermitian matrix $\bar{A}^tA$ are non-negative.

Proof

The eigenvalues of $\bar{A}^tA$ are all real by (4.3.6).

Let λ be an eigenvalue of $\bar{A}^tA$ so that $\bar{A}^tAv = \lambda v$ for some non-zero vector $v \in \mathbb{C}^n$. Then $\bar{v}^t\bar{A}^tAv = \lambda\bar{v}^tv$, i.e. $\|Av\|^2 = \lambda\|v\|^2$ where $\| \;\; \|$ denotes the euclidean norm. Hence $\lambda \ge 0$ since $\|v\| > 0$.

8.3.3 **Definition**

Let A be an $m{\times}n$ matrix with complex entries.The *singular values* of A are $\sigma_1, \sigma_2, \ldots, \sigma_n$ where $\sigma_1 \ge \sigma_2 \ge \ldots \ge \sigma_n \ge 0$ and $\sigma_1^2, \sigma_2^2, \ldots, \sigma_n^2$ are the eigenvalues of $\bar{A}^tA$.

8.3.4 **Lemma**

Let A be an $m{\times}n$ matrix with complex entries.

Then rank A = rank $\bar{A}^tA$ = rank $A\bar{A}^t$.

Proof

Recall from (iii) of (2.6.13) that rank $\bar{A}^tA \le$ rank A.

To obtain the reverse inequality note first that $\bar{A}^tAv = 0$ implies that $\bar{v}^t\bar{A}^tAv = 0$, i.e. $\| \, Av \, \|^2 = 0$. The properties of norms now imply that $Av = 0$. Thus the kernel of $\bar{A}^tA$ is contained in the kernel of A. The fact that rank A $\le$ rank $\bar{A}^tA$ now follows from the rank-nullity theorem (2.6.5).

To show that rank A = rank $A\bar{A}^t$ we replace A by $\bar{A}^t$ in the above argument and recall from (i) of (2.6.13) that rank A = rank $\bar{A}^t$.

8.3.5 **Remark**

If rank A = r where r ≤ min (m,n) then the first r of the singular values of A will be positive and all of the others will be zero.

8.3.6 **Exercise**

Let $A \in M_n\mathbb{C}$. Show that λ is a non-zero eigenvalue of $\bar{A}^tA$ if and only if λ is a non-zero eigenvalue of $A\bar{A}^t$.

8.3.7 **Remark**

It follows from (8.3.6) that A and $\bar{A}^t$ have the same set of non-zero singular values but for m ≠ n one of them will have more zeros occurring as singular values.

8.3.8 **Exercise**

Let $A \in M_n\mathbb{C}$ be hermitian with eigenvalues $\lambda_1,\lambda_2,...,\lambda_n$. Show that the singular values of A are $|\lambda_1|,|\lambda_2|,...,|\lambda_n|$.

8.3.9 **Theorem** **(The Singular Value Decomposition)**

Let A be an m×n matrix with complex entries. Let $\sigma_1,\sigma_2,...,\sigma_r$ be the non-zero singular values of A where r = rank A and the σ_i are written in decreasing order, i.e. $\sigma_1 \geq \sigma_2 \geq .. \geq \sigma_r > 0$. Let D be the r×r matrix with entries $\sigma_1,\sigma_2,...,\sigma_r$ on the diagonal and zero elsewhere. Let Σ be the m×n matrix with D as the top left-hand r×r block and zero everywhere else. Then there exists a unitary m×m matrix U and a unitary n×n matrix V such that $\bar{U}^tAV = \Sigma$.

Proof

First we will prove by induction that there exist unitary matrices U and V of size m and n respectively such that $\bar{U}^tAV = S$ where S is a matrix with a real non-negative diagonal block in the top left-hand corner and zero everywhere else. We will then show that S necessarily equals the matrix Σ.

Our induction will be on min(m,n). Suppose min(m,n) = 1 so that A is either m×1 or 1×n. If A is m×1 then $A = (a_i)$ is a column vector. Let U be a unitary m×m matrix whose first column is the vector (a_i) normalized, i.e. the first column is the unit vector $(a_i/\|A\|)$ where $\| \ \|$ denotes the euclidean norm.

(This is always possible by (b) of (2.2.13) and (2.7.7) !)

Let V be the 1×1 matrix with entry 1. Then $\bar{U}^tAV$ is the 1×n matrix $\begin{pmatrix} \|A\| \\ 0 \\ \vdots \\ 0 \end{pmatrix}$. Similarly if A is a 1×n matrix, i.e. a row vector (a_i) of length n, we take U to be the 1×1 matrix with entry 1 and V to be a unitary n×n matrix with (a_i) normalized as its first column. Then $\bar{U}^tAV = (\|A\|, 0, \ldots, 0)$. Thus the theorem is true when min (m,n) = 1.

Now assume the theorem is true for all m×n matrices with min (m,n) < k. Let A be an m×n matrix and let $\sigma = \|A\|$ be the spectral norm of A, i.e. the operator norm corresponding to the euclidean norm on $\mathbb{C}^n$ and $\mathbb{C}^m$. Choose x in $\mathbb{C}^n$ such that $\|x\| = 1$ and $\|Ax\| = \sigma$. (This is always possible because of the definition of operator norm given in (3.3.9).)

Let $y \in \mathbb{C}^n$ be defined by $y = \sigma^{-1}Ax$. (We may assume $A \neq 0$ so that $\sigma \neq 0$ since for A = 0 the theorem is trivially true.) Then $Ax = \sigma y$ and $\|y\| = 1$. Now let U be a unitary m×m matrix with y as its first column and let V be a unitary n×n matrix with x as its first column.

Then $\bar{U}^tAV = \begin{pmatrix} \sigma & \bar{w}^t \\ 0 & \\ \vdots & C \\ 0 & \end{pmatrix}$ where w is an (n-1)×1 matrix and C is an (m-1)×(n-1) matrix.

Observe that $(\bar{U}^t AV)\begin{pmatrix}\sigma\\ w\end{pmatrix} = \begin{pmatrix}\sigma^2 + \bar{w}^t w\\ Cw\end{pmatrix}$ and this gives the inequality

$$\| (\bar{U}^t AV)\begin{pmatrix}\sigma\\ w\end{pmatrix} \| \geq \sigma^2 + \bar{w}^t w.$$

Now $\|\bar{U}^t AV\| \leq \|A\|$ since $\|U\| = \|V\| = 1$ for U and V unitary and $\| \ \|$ satisfies property (*) of (3.3.6). Also, from the definition of operator norm, we see that $\| (\bar{U}^t AV)\begin{pmatrix}\sigma\\ w\end{pmatrix} \| \leq \| \bar{U}^t AV \| \ \| \begin{pmatrix}\sigma\\ w\end{pmatrix} \|$ and hence we find that $\sigma^2 + \bar{w}^t w \leq \|A\|^2$. However $\sigma = \|A\|$ and hence the last inequality implies that $w = 0$. We can now apply the inductive assumption to the $(m-1)\times(n-1)$ matrix C and this completes the induction argument.

For the last part of the proof suppose that there exists a unitary $m\times m$ matrix U and a unitary $n\times n$ matrix V such that $\bar{U}^t AV = S$ where S has zeros everywhere except for the top left-hand $r\times r$ block which is diagonal with non-negative entries $\gamma_1, \gamma_2, \ldots, \gamma_r$ on the diagonal. We may assume that $\gamma_1 \geq \gamma_2 \geq \ldots \geq \gamma_r$ since if not we could permute the columns of U and V to achieve this.

The equation $\bar{U}^t AV = S$ means that $AV = US$ so that $Av_i = \gamma_i u_i$ for each $i = 1,2,\ldots,r$. (Here u_i, v_i denote the columns of U,V respectively.) Also $\bar{U}^t A = S\bar{V}^t$ implies that $\bar{u}_i^t A = \gamma_i \bar{v}_i^t$ for each $i = 1.2,\ldots,r$ and thus $\bar{A}^t u_i = \gamma_i v_i$. Hence $\bar{A}^t A v_i = \gamma_i^2 v_i$ for each i so that $\gamma_1^2, \gamma_2^2, \ldots, \gamma_r^2$ are the non-zero eigenvalues of $\bar{A}^t A$. This shows that $\gamma_i = \sigma_i$ for each $i = 1,2,\ldots,r$ and the proof is complete.

8.3.10 <u>Note</u>

The columns of V are a full orthonormal set of eigenvectors of the $n\times n$ matrix $\bar{A}^t A$. The columns of U are a full orthonormal set of eigenvectors of the $m\times m$ matrix $A\bar{A}^t$. If all of the entries of A are real then U and V may be chosen to be real orthogonal matrices.

8.3.11 **Remark**

The induction proof in (8.3.9) may seem somewhat reminiscent of the proof of Schur's theorem (6.1.3). There is a good reason for this. The special case of (8.3.9) when m = n and A is normal will give U = V since $A\bar{A}^t = \bar{A}^t A$. Thus (8.3.9) may be viewed as a generalization of Schur's theorem (6.1.3) for normal matrices.

8.3.12 **Remark**

In the proof of (8.3.9) notice that $\sigma = \|A\|$ appears as one of the singular values of A, $\| \ \|$ denoting the spectral norm. If λ is an eigenvalue of $\bar{A}^t A$ then $\bar{A}^t A v = \lambda v$ for some $v \neq 0$. Hence we have that $\bar{v}^t \bar{A}^t A v = \lambda \bar{v}^t v$ so that $\lambda = \dfrac{\|Av\|}{\|v\|}$. It follows that $\|A\|$ is the largest singular value of A. Thus (8.3.9) provides another proof of (6.8.7), valid for rectangular matrices as well as for square matrices.

8.3.13 **Example**

Find the singular value decomposition of $A = \begin{pmatrix} 2 & -1 \\ -2 & 1 \\ 4 & -2 \end{pmatrix}$.

It is easy to check that $A^t A = \begin{pmatrix} 24 & -12 \\ -12 & 6 \end{pmatrix}$ and that this 2x2 matrix has characteristic polynomial $\lambda^2 - 30\lambda$. Thus $\lambda = 30$, $\lambda = 0$ are the two eigenvalues of $A^t A$ and so $\sigma = \sqrt{30}$ is the only non-zero singular value of A.

The singular value decomposition of A will be the matrix $\Sigma = \begin{pmatrix} \sqrt{30} & 0 \\ 0 & 0 \\ 0 & 0 \end{pmatrix}$. We can determine the unitary matrices U and V such that $\bar{U}^t A V = \Sigma$ in the following way, making use of (8.3.10). Note that U and V can be chosen to be real orthogonal since all of the entries of A are real.

For $\lambda = 30$ an easy calculation shows that $\begin{pmatrix} 2/\sqrt{5} \\ -1/\sqrt{5} \end{pmatrix}$ is a unit eigenvector of A^tA and for $\lambda = 0$ that $\begin{pmatrix} 1/\sqrt{5} \\ 2/\sqrt{5} \end{pmatrix}$ is a unit eigenvector. Thus we may take

$$V = \begin{pmatrix} 2/\sqrt{5} & 1/\sqrt{5} \\ -1/\sqrt{5} & 2/\sqrt{5} \end{pmatrix}.$$

Now $AA^t = \begin{pmatrix} 5 & -5 & 10 \\ -5 & 5 & -10 \\ 10 & -10 & 20 \end{pmatrix}$ which has eigenvalues 30, 0, 0. For $\lambda = 30$ the vector $\begin{pmatrix} 1/\sqrt{6} \\ -1/\sqrt{6} \\ 2/\sqrt{6} \end{pmatrix}$ is a unit eigenvector. For $\lambda = 0$ the eigenvectors are the non-zero solutions of $x - y + 2z = 0$. To make U orthogonal we can choose $\begin{pmatrix} 1/\sqrt{2} \\ 1/\sqrt{2} \\ 0 \end{pmatrix}$ and $\begin{pmatrix} 1/\sqrt{3} \\ -1/\sqrt{3} \\ -1/\sqrt{3} \end{pmatrix}$ as the second and third columns of U. Thus we may take

$$U = \begin{pmatrix} 1/\sqrt{6} & 1/\sqrt{2} & 1/\sqrt{3} \\ -1/\sqrt{6} & 1/\sqrt{2} & -1/\sqrt{3} \\ 2/\sqrt{6} & 0 & -1/\sqrt{3} \end{pmatrix}.$$

The reader should check that $U^tAV = \Sigma$.

8.3.14. Numerical computation and singular values

The singular value decomposition is of great importance in numerical matrix computation, in particular for problems of "near rank-deficiency" which we discuss below. An extension of the QR-algorithm (8.2.6) may be used for the numerical computation of singular values.

We show now that the singular values of any matrix are perfectly-conditioned, i.e. the matrix A is near to the matrix B if and only if the singular value decomposition of A is near to the singular value decomposition of B. (This is in contrast to the situation for the eigenvalues of a matrix which we saw in (7.1) to be ill-conditioned in general.) We require a few preliminary results.

8.3.15 **Exercise**

Let A be any m×n matrix and let $\tilde{A} = \begin{pmatrix} 0 & A \\ \bar{A}^t & 0 \end{pmatrix}$ which will be a hermitian (m + n)×(m + n) matrix.

Show that $\| \tilde{A} \| = \| A \|$ where $\| \ \|$ denotes the spectral norm.

[*Hint* - use (8.3.12)]

8.3.16 **Exercise**

Let A and $\tilde{A}$ be as in (8.3.15). Let $\bar{U}^t AV = \Sigma$ be the singular value decomposition of A as described in (8.3.9) where $\Sigma = \begin{pmatrix} D & 0 \\ 0 & 0 \end{pmatrix}$, D being diagonal with the non-zero singular values $\sigma_1, \sigma_2, \ldots, \sigma_r$ of A on the diagonal.

Show that $\tilde{A}$ is unitarily similar to the matrix $\begin{pmatrix} D & 0 & 0 \\ 0 & -D & 0 \\ 0 & 0 & 0 \end{pmatrix}$.

[*Hint* - assume that $m \geq n$ and write the m×m unitary matrix U in the form (U_1, U_2) where U_1 is m×n and U_2 is m×(m-n).

Then writing $P = \begin{pmatrix} (1/\sqrt{2})U_1 & -(1/\sqrt{2})U_1 & U_2 \\ (1/\sqrt{2})V & (1/\sqrt{2})V & 0 \end{pmatrix}$, show that P is unitary and that

$\bar{P}^t \tilde{A} P = \begin{pmatrix} D & 0 & 0 \\ 0 & -D & 0 \\ 0 & 0 & 0 \end{pmatrix}$. If $m < n$ a similar argument works.]

8.3.17 **Exercise**

Deduce from (8.3.16) that $\sigma_1, \sigma_2, \ldots, \sigma_r$ are the non-zero singular values of A if and only if $\sigma_1, \sigma_2, \ldots, \sigma_r, -\sigma_1, -\sigma_2, \ldots, -\sigma_r$ are the non-zero singular values of $\tilde{A}$.

8.3.18 **Proposition**

Let A be an m×n matrix with entries in $\mathbb{C}$. Let A be perturbed into a new matrix A + δA. Let $\sigma_1 \le \sigma_2 \le \ldots \le \sigma_n$ be the singular values of A and let $\mu_1 \le \mu_2 \le \ldots \le \mu_n$ be the singular values of A + δA. Then $|\sigma_i - \mu_i| \le \| \delta A \|$ for each i = 1,2,. . .,n, where $\| \ \|$ denotes the spectral norm, i.e. the singular values of A are perfectly conditioned.

Proof

Let B be another m×n matrix and apply (6.9.2) to the hermitian matrices $\tilde{A}$ and $\tilde{B}$ to obtain the inequality

$$|\lambda_i(\tilde{A} + \tilde{B}) - \lambda_i(\tilde{B})| \le \lambda_{m+n}(\tilde{B})$$

for each i = 1,2,. . . ,m + n, where the λ_i denote the eigenvalues written in ascending order. Now $\lambda_{m+n}(\tilde{B}) \le \|\tilde{B}\|$ by (4.4.1) and so, putting B = δA and using (8.3.15) and (8.3.17), we obtain the result.

8.3.19 **Rank-deficiency**

An m×n matrix A is said to be *rank-deficient* if rank A < min(m,n). For square matrices rank-deficient means singular.

If the entries of a matrix are only known approximately and we wish to know whether or not the matrix is of maximal rank (i.e. not rank-deficient) then we should ask whether or not the matrix is near to a rank-deficient matrix. The singular values are very helpful in answering this question.

8.3.20 **Theorem**

Let A be an m×n matrix with complex entries and let r = rank A.

Let $\sigma_1 \ge \sigma_2 \ge \ldots \ge \sigma_r$ be the non-zero singular values of A. Suppose k is a natural number with k < r.

Then $\min_B \|A - B\| = \sigma_{k+1}$ where this minimum is taken over all m×n matrices B of rank k and $\| \ \|$ is the spectral norm.

(i.e. any m×n matrix of rank k is of distance at least σ_{k+1} away from A.)

Proof

Let $\{u_1,u_2,...,u_m\}$ and $\{v_1,v_2,...,v_n\}$ be orthonormal sets of eigenvectors of $A\bar{A}^t$ and $\bar{A}^tA$ respectively as in the proof of (8.3.9). Recall that $Av_i = \sigma_i u_i$ for each $i = 1,2,...,r$.

Let B be any matrix of rank k. Then the kernel of B has dimension n-k. Let W be the (k+1)-dimensional subspace of $\mathbb{C}^n$ spanned by $\{v_1,v_2,...,v_{k+1}\}$. By dimension considerations the intersection of W with the kernel of B is non-zero. Choose a vector x in this intersection for which $\|x\| = 1$ using the euclidean norm. Writing $x = \sum_{i=1}^{k+1} \alpha_i v_i$ we see, by orthonormality of the vectors v_i, that $\alpha_i = \bar{v}_i^t x$ for each $i = 1,2,...,k+1$. Also $\|x\| = 1$ implies that $\sum_{i=1}^{k+1} |\bar{v}_i^t x|^2 = 1$.

Since $Bx = 0$, $(A - B)x = Ax = \sum_{i=1}^{k+1} \bar{v}_i^t x A v_i = \sum_{i=1}^{k+1} \bar{v}_i^t x \sigma_i u_i$.

Hence $\|(A - B)x\|^2 = \sum_{i=1}^{k+1} |\bar{v}_i^t x|^2 \sigma_i^2$ because the u_i are orthonormal.

Also $\|(A - B)x\| \leq \|A - B\| \, \|x\| = \|A - B\|$ since $\|x\| = 1$. This yields that $\|A - B\|^2 \geq \sum_{i=1}^{k+1} |\bar{v}_i^t x|^2 \sigma_i^2 \geq \sigma_{k+1}^2$ because of the ordering of the σ_i and the fact that $\sum_{i=1}^{k+1} |\bar{v}_i^t x|^2 = 1$. This completes the proof.

8.3.21 **Corollary**

Let A have maximal rank, i.e. $r = \min(m,n)$. Then any m×n matrix which is rank-deficient is of distance at least σ_r away from A, σ_r being the smallest non-zero singular value.

Proof

Immediate from (8.3.20).

8.3.22 **Lemma**

Let $A = (a_{ij})$ and $B = (b_{ij})$ be m×n matrices with complex entries. Suppose that $B = \bar{P}^tAQ$ where P is a unitary m×m matrix and Q is a unitary n×n matrix. Then $\sum_{i,j} |a_{ij}|^2 = \sum_{i,j} |b_{ij}|^2$, i.e. A and B have the same euclidean norm.

Proof

The proof of lemma (7.3.1) for square matrices which are unitarily equivalent goes through in this more general situation.

8.3.23 **Theorem**

Let A be an m×n matrix with complex entries and let $\| \ \|$ be any norm on the space of m×n matrices. Given any $\varepsilon > 0$ there exists a matrix A_1 with distinct singular values and for which $\|A - A_1\| < \varepsilon$.

Proof

We prove the result using the euclidean norm. The result follows for any other choice of norm because, by (3.2.11), all norms on a finite dimensional real or complex vector space are equivalent.

Let $\bar{U}^tAV = \Sigma$ be the singular value decomposition of A. It is easy to see that there exists an m×n matrix Σ_1 with distinct singular values and with $\|\Sigma - \Sigma_1\| < \varepsilon$. (We only need to adjust the (1,1),(2,2),. . . .,(r,r) entries of Σ to make them all different. Here $r = \min(m,n)$.)

Let $A_1 = U\Sigma_1\bar{V}^t$ so that Σ_1 is the singular value decomposition of A_1. Then $\|A - A_1\| = \|U(\Sigma - \Sigma_1)\bar{V}^t\| = \|\Sigma - \Sigma_1\|$ using (8.3.22). Thus $\|A - A_1\| < \varepsilon$ and the proof is complete.

8.3.24 **Remark**

We saw in (7.3.2) that any square matrix can be approximated by a matrix with distinct eigenvalues. Theorem (8.3.23) gives a similar result for singular values.

8.3.25 Min-max theorem for singular values

Let A be an m×n matrix with complex entries. Let the non-zero singular values of A be $\sigma_1 \le \sigma_2 \le \ldots \le \sigma_r$. Then for any k, $1 \le k \le r$, $\sigma_k = \min\left(\max \frac{\|Av\|}{\|v\|}\right)$ where $\| \ \|$ denotes the euclidean norm and where the min and max are taken over exactly the same sets as in the Courant-Fisher min-max theorem (6.9.1).

Proof

Applying (6.9.1) to the matrix $\bar{A}^tA$ yields that $\sigma_k^2 = \min(\max \rho_{\bar{A}^tA}(v))$.

The result follows from the fact that $\rho_{\bar{A}^tA}(v) = \frac{\bar{v}^t\bar{A}^tAv}{\bar{v}^tv} = \frac{\|Av\|^2}{\|v\|^2}$.

Problems 8C

1. Find the singular value decomposition of each of the following;

(i) $\begin{pmatrix} 2 & -2 \\ 1 & 1 \\ 2 & 2 \end{pmatrix}$, (ii) $\begin{pmatrix} 2 & 1 & -1 \\ -2 & -1 & 1 \end{pmatrix}$, (iii) $\begin{pmatrix} 1 & 2 & 2 & 3 \\ 2 & 0 & -1 & 0 \\ 2 & -5 & 4 & 0 \end{pmatrix}$.

2. Let A be an m×n matrix with rank r and whose non-zero singular values are $\sigma_1, \sigma_2, \ldots, \sigma_r$. Show that the euclidean norm of A is given by

$$\|A\| = (\sigma_1^2 + \sigma_2^2 + \ldots + \sigma_r^2)^{1/2}.$$

3. Use the singular value decomposition to show that the set of all maximal rank m×n matrices is (i) an open subset, (ii) a dense subset, of the set of all m×n matrices.

[*Hint* - use the method of proof of (8.3.23) to show (i) if A has maximal rank and B is close to A then B also has maximal rank, (ii) If A is any m×n matrix then there exists a matrix B of maximal rank and close to A.]

4. Let $\sigma_i(A)$, $i = 1,2,\ldots,r$ be the non-zero singular values of the m×n matrix A. On the space of all m×n matrices we define the *trace norm*, denoted $\| \ \|_{tr}$ by $\|A\|_{tr} = \sum_{i=1}^{r} \sigma_i(A)$.

Show that $\| \ \|_{tr}$ is a norm and that it is unitarily invariant in the sense that $\| UAV \|_{tr} = \| A \|_{tr}$ for any m×m unitary matrix U and any unitary n×n matrix V.

5. Show that $\begin{pmatrix} 1 & 1 \\ 0 & 1 \end{pmatrix}$ and $\begin{pmatrix} 2 & -1 \\ 1 & 0 \end{pmatrix}$ are similar matrices but that they do not have the same singular values.

6. Let A be an n×n matrix which is not unitary and let $\bar{U}^t AV = \Sigma$ be the singular value decomposition of A.

Show that, using the euclidean norm, the unitary matrix $B = U\bar{V}^t$ is the unitary matrix closest to A, (i.e. show that if P is a unitary n×n matrix then $\|B - A\| \le \|P - A\|$.)

[*Hint* - observe that $\|P - A\| = \|P - U\Sigma\bar{V}^t\| = \|\bar{U}^t PV - \Sigma\|$ by (8.3.22), and so $\bar{U}^t BV$ must be the unitary matrix closest to Σ. Then show that $\|P - \Sigma\|^2 \ge \|I - \Sigma\|^2$ for any unitary matrix P, i.e. that I is the unitary matrix closest to Σ.]

8.4 Generalized inverses of matrices

8.4.1 Definition

Let A be an m×n matrix with complex entries.

The *generalized inverse* of A is the unique n×m matrix B which satisfies each of the following four conditions;

(i) AB is hermitian, i.e. $AB = (\overline{AB})^t$.

(ii) BA is hermitian, i.e. $BA = (\overline{BA})^t$.

(iii) $ABA = A$.

(iv) $BAB = B$.

To show that the above definition is meaningful we must show that there exists an $n \times m$ matrix B satisfying (i), (ii), (iii), and (iv) and that this matrix B is the only one which satisfies the four conditions.

8.4.2 **Proposition**

Let A be an $m \times n$ matrix with complex entries. Then there exists a unique $n \times m$ matrix B satisfying (i), (ii), (iii), and (iv).

Proof

(Existence)

Let Σ be the singular value decomposition of A as in (8.3.9) so that $\bar{U}^t AV = \Sigma$ for unitary matrices U and V. Recall that Σ has the $r \times r$ diagonal block D in the top left corner, the diagonal entries of D being the non-zero singular values of A and r being the rank of A. Let Σ_1 be the $n \times m$ matrix with the $r \times r$ block D^{-1} in the top left corner and zero everywhere else. Note that $\Sigma\Sigma_1 = \Sigma_1^t\Sigma^t$, $\Sigma_1\Sigma = \Sigma^t\Sigma_1^t$, $\Sigma\Sigma_1\Sigma = \Sigma$, and $\Sigma_1\Sigma\Sigma_1 = \Sigma_1$. Let $B = V\Sigma_1\bar{U}^t$. We will show that B satisfies (i), (ii), (ii), and (iv).

(i) $$\begin{aligned} AB &= U\Sigma\bar{V}^t V\Sigma_1\bar{U}^t \\ &= U\Sigma\Sigma_1\bar{U}^t \\ &= U\Sigma_1^t\Sigma^t\bar{U}^t \\ &= U\Sigma_1^t\bar{V}^t V\Sigma^t\bar{U}^t \\ &= \bar{B}^t\bar{A}^t \\ &= \overline{(AB)}^t \end{aligned}$$

(ii) $$\begin{aligned} BA &= V\Sigma_1\bar{U}^t U\Sigma\bar{V}^t \\ &= V\Sigma_1\Sigma\bar{V}^t \\ &= V\Sigma^t\Sigma_1^t\bar{V}^t \\ &= V\Sigma^t\bar{U}^t U\Sigma_1^t\bar{V}^t \\ &= \bar{A}^t\bar{B}^t \\ &= (\overline{BA})^t \end{aligned}$$

(iii) $ABA = U\Sigma\bar{V}^tV\Sigma_1\bar{U}^tU\Sigma\bar{V}^t$

$= U\Sigma\Sigma_1\Sigma\bar{V}^t$

$= U\Sigma\bar{V}^t$

$= A.$

(iv) $BAB = V\Sigma_1\bar{U}^tU\Sigma\bar{V}^tV\Sigma_1\bar{U}^t$

$= V\Sigma_1\Sigma\Sigma_1\bar{U}^t$

$= V\Sigma_1\bar{U}^t$

$= B.$

(Uniqueness)

Suppose that the n×m matrices B_1 and B_2 each satisfy (i), (ii), (iii), and (iv).

Then $\bar{B}_1^t - \bar{B}_2^t = \bar{B}_1^t\bar{A}^t\bar{B}_1^t - \bar{B}_2^t\bar{A}^t\bar{B}_2^t = AB_1\bar{B}_1^t - AB_2\bar{B}_2^t$ since B_1 and B_2 each satisfy (i). Hence Im $(\bar{B}_1^t - \bar{B}_2^t)$ is contained in Im A.

Also $A\bar{A}^t(\bar{B}_1^t - \bar{B}_2^t) = AB_1A - AB_2A = A - A = 0$ after using (i) and (iii). Hence Im $(\bar{B}_1^t - \bar{B}_2^t)$ is contained in Ker $A\bar{A}^t$.

Now $A\bar{A}^tv = 0$ implies that $B_1A\bar{A}^tv = 0$. Using (ii) and (iii) we see that $B_1A\bar{A}^t = \bar{A}^t\bar{B}_1^t\bar{A}^t = \bar{A}^t$. Thus $A\bar{A}^tv = 0$ implies that $\bar{A}^tv = 0$. This shows that Ker $A\bar{A}^t$ = Ker $\bar{A}^t$. Thus we know that Im $(\bar{B}_1^t - \bar{B}_2^t)$ is contained in (Im A)$\cap$(Ker $\bar{A}^t$). If $z \in$ (Im A)$\cap$(Ker $\bar{A}^t$) then $z = Aw$ for some w and $\bar{A}^tAw = 0$. This yields that $\bar{w}^t\bar{A}^tAw = 0$, i.e. $\|Aw\|^2 = 0$. The properties of norm ensure that $z = Aw = 0$.

We have now shown that Im $(\bar{B}_1^t - \bar{B}_2^t) = 0$ from which we see that $\bar{B}_1^t = \bar{B}_2^t$. The result $B_1 = B_2$ follows at once by taking conjugate transposes.

8.4.3 **Remark on terminology**

The generalized inverse defined in (8.4.1) is often called the *Moore-Penrose generalized inverse*. Some authors refer to any matrix B satisfying (iii) and (iv) of (8.4.1) as a generalized inverse. Generalized inverses are useful and illuminating in a number of different areas including differential and difference equations, statistics, numerical analysis, and electrical networks. In this book we will confine ourselves to one application, the use of the Moore-Penrose generalized inverse to formulate a very neat solution of the least squares problem. The reader who wishes to see other applications should consult [C] and [N].

8.4.4 **Notation**

We will write A^+ for the Moore-Penrose generalized inverse of the matrix A.

8.4.5 **Example**

Find A^+ for the matrix $A = \begin{pmatrix} 2 & -1 \\ -2 & 1 \\ 4 & -2 \end{pmatrix}$ of example (8.3.13).

Using (8.4.2) the Moore-Penrose generalized inverse $A^+ = V\Sigma_1 U^t$. We saw in (8.3.13) that $\Sigma = \begin{pmatrix} \sqrt{30} & 0 \\ 0 & 0 \\ 0 & 0 \end{pmatrix}$ so that $\Sigma_1 = \begin{pmatrix} 1/\sqrt{30} & 0 & 0 \\ 0 & 0 & 0 \end{pmatrix}$. We also saw that

$$U = \begin{pmatrix} 1/\sqrt{6} & 1/\sqrt{2} & 1/\sqrt{3} \\ -1/\sqrt{6} & 1/\sqrt{2} & -1/\sqrt{3} \\ 2/\sqrt{6} & 0 & -1/\sqrt{3} \end{pmatrix} \text{ and } V = \begin{pmatrix} 2/\sqrt{5} & 1/\sqrt{5} \\ -1/\sqrt{5} & 2/\sqrt{5} \end{pmatrix}.$$

$$\text{Hence } A^+ = \begin{pmatrix} 2/\sqrt{5} & 1/\sqrt{5} \\ -1/\sqrt{5} & 2/\sqrt{5} \end{pmatrix} \begin{pmatrix} 1/\sqrt{30} & 0 & 0 \\ 0 & 0 & 0 \end{pmatrix} \begin{pmatrix} 1/\sqrt{6} & -1/\sqrt{6} & 2/\sqrt{6} \\ 1/\sqrt{2} & 1/\sqrt{2} & 0 \\ 1/\sqrt{3} & -1/\sqrt{3} & -1/\sqrt{3} \end{pmatrix}$$

$$= \begin{pmatrix} 1/15 & -1/15 & 2/15 \\ -1/30 & 1/30 & -1/15 \end{pmatrix}.$$

As an exercise the reader may check that conditions (i), (ii), (iii), and (iv) of definition (8.4.1) are satisfied by A^+.

Problems 8D

1. Determine the Moore-Penrose generalized inverse of each of the three matrices in problem 1 of Problems 8C.

2. Let A be a non-singular n×n matrix. Show that $A^+ = A^{-1}$.

If A is an m×n matrix for which $\bar{A}^tA$ is non-singular show that $A^+ = (\bar{A}^tA)^{-1}\bar{A}^t$.

3. Let A be an m×n matrix and suppose $\bar{A}^tA$ is singular. Show that $\bar{A}^tA + \varepsilon I$ is non-singular for $\varepsilon > 0$ provided that ε is small enough. Deduce that

$$A^+ = \lim_{\varepsilon \to 0} (\bar{A}^tA + \varepsilon I)^{-1}\bar{A}^t.$$

[*Hint* - for the last part let $B_\varepsilon = (\bar{A}^tA + \varepsilon I)^{-1}\bar{A}^t$, let $B = \lim_{\varepsilon \to 0} B_\varepsilon$, and examine AB_ε, $B_\varepsilon A$, $AB_\varepsilon A$, and $B_\varepsilon AB_\varepsilon$ to show that B satisfies (i), (ii), (iii), and (iv) in the definition of A^+.]

8.5 **Least squares problems**

8.5.1 **Statement of the problem**

Let A be an m×n matrix with real entries, let b be a column vector in $\mathbb{R}^m$, and let x be a column vector in $\mathbb{R}^n$. The matrix equation $Ax = b$ represents a system of m simultaneous linear equations in n unknowns $x_1, x_2, \ldots, x_n$ where $x = (x_i)$. We have seen in chapter 1 that there are three possibilities for the solution set of $Ax = b$, namely;

(i) there is a unique solution,

(ii) there is an infinite set of solutions,

(iii) there is no solution at all.

If the last of these three possibilities occurs we may ask whether there is an $x \in \mathbb{R}^n$ which, in some sense, is nearest to being a solution.

Since whenever $m \leq n$ the solution set satisfies either (i) or (ii) we will assume that $m > n$ for the remainder of this section.

Problem

Find $x \in \mathbb{R}^n$ for which $\| Ax - b \|$ is as small as possible, $\| \;\|$ denoting *the euclidean norm on* $\mathbb{R}^n$.

This problem is known as a *least squares problem* because the quantity $\| Ax - b \|^2$ is a sum of squares and it is clear that minimizing $\| Ax - b \|^2$ is equivalent to minimizing $\| Ax - b \|$. At the moment it may not be clear whether this problem necessarily has any solution at all and, if it does have a solution, whether the solution is unique. This will be answered in (8.5.8).

8.5.2 **Euclidean and other norms**

The reader may wonder why we choose the euclidean norm rather than any other norm in formulating the problem in (8.5.1). There are a few reasons for this ;

(i) The euclidean norm gives the usual notion of distance in $\mathbb{R}^n$ and so geometric arguments may be used.

(ii) $\| Ax - b \|^2$ is a continuously differentiable function for the euclidean norm $\| \ \|$.

(iii) The euclidean norm on $\mathbb{R}^n$ is unaltered by orthogonal transformations, i.e. $\| Ax - b \| = \| QAx - Qb \|$ for any orthogonal matrix Q. An appropriate choice of Q may reduce QA to a very simple form.

We most certainly could formulate the problem in (8.5.1) using some other norm, e.g. the cartesian norm. There is no reason to expect that the answer to the problem would be the same as that for the euclidean norm. It is true that all norms on $\mathbb{R}^n$ are equivalent in the sense described in chapter 3 and in particular equivalent norms will be the same for continuity and convergence properties. However actual magnitudes are different for different norms so that a vector $x \in \mathbb{R}^n$ making the cartesian norm $\| Ax - b \|$ minimal need not make the euclidean norm $\| Ax - b \|$ minimal, and vice versa. We use the euclidean norm for the reasons stated earlier.

8.5.3 Example

The classical case of the least squares problem is that of finding "the straight line of best fit" for a given set of points in the plane.

For example suppose we are given a set of points (x_i,y_i), $i = 1,2,\ldots,m$ in the plane and we believe there is a linear relationship between y and x, i.e. $y = \alpha x + \beta$ for some constants α and β. (A physical example would be when y represents the velocity at time x of a particle travelling in a straight line with constant acceleration.) The "line of best fit by the method of least squares" to the given data will be the line $y = \alpha x + \beta$ where α and β are chosen so as to minimize $\sum_{i=1}^{m} (y_i - \alpha x_i - \beta)^2$.
This is indeed a special case of the problem posed in (8.5.1) with n = 2,

$$A = \begin{pmatrix} 1 & x_1 \\ 1 & x_2 \\ \vdots & \vdots \\ 1 & x_m \end{pmatrix},\ b = \begin{pmatrix} y_1 \\ y_2 \\ \vdots \\ y_m \end{pmatrix},\ \text{and } x = \begin{pmatrix} \beta \\ \alpha \end{pmatrix}.$$

8.5.4 The normal equations

Given a system of simultaneous linear equations $Ax = b$ as described in (8.5.1) the set of equations $A^tAx = A^tb$ is called *the set of normal equations* corresponding to the system $Ax = b$.

The normal equations lead to a good theoretical way of solving the least squares problem.

8.5.5 Lemma

The solution set of the least squares problem of minimizing $\| Ax - b \|$ coincides with the solution set of the system of linear equations

$$A^tAx = A^tb.$$

Proof

Let $z \in \mathbb{R}^n$ be a solution of the least squares problem. Then, from geometrical considerations, $Az - b$ must be orthogonal to every vector in Im A. The point in Im A closest to b must be the perpendicular projection of b onto Im A.

Hence Ay is orthogonal to Az - b for all vectors $y \in \mathbb{R}^n$. This implies that $\bar{y}^t A^t(Az - b) = 0$ for all $y \in \mathbb{R}^n$. Hence $A^t(Az - b) = 0$ so that z satisfies the normal equations.

Conversely if z satisfies the normal equations then, by reversing the above argument, we see that Az - b is orthogonal to Im A. Thus z is a solution of the least squares problem.

8.5.6 **Comment**

We have given a geometric proof of (8.5.5). If this kind of argument is not to the reader's taste then an alternative proof, via calculus, is contained in problem 3 of Problems 8E.

8.5.7 **Computation of least squares solutions**

Theoretically (8.5.5) gives a good way to solve a least squares problem by solving $A^tAx = A^tb$. From the point of view of numerical computation the normal equations are not so good. This is because the condition number $c(A^tA) = c(A)^2$ so that errors may be magnified on passing to the normal equations.

8.5.8 **Theorem**

Let A be an m×n matrix with real entries and with m > n. Let $x \in \mathbb{R}^n$, $b \in \mathbb{R}^m$. Then the least squares problem of minimizing $\| Ax - b \|$ always has a solution. It has a unique solution if and only if rank A = n. This unique solution is given by $x = A^+b$ where A^+ is the Moore-Penrose generalized inverse of A. For rank A < n there is an infinite set of solutions of the least squares problem. In this infinite solution set there is a unique element x of minimal euclidean norm and this element is given by $x = A^+b$.

Proof

We first show that $x = A^+b$ is always a solution of the least squares problem. By (8.5.5) it suffices to show that $x = A^+b$ satisfies the normal equations $A^tAx = A^tb$.

We have $A = U\Sigma V^t$ by (8.3.9) and $A^+ = V\Sigma_1U^t$ by (8.4.2) where V, U, Σ, Σ_1 are as in (8.3.9) and (8.4.2). (Note that U and V can be chosen to be real orthogonal since all of the entries of A are real.) Hence when $x = A^+b$ we find that

$$\begin{aligned} A^tAx &= A^tAA^+b \\ &= V\Sigma^tU^tU\Sigma V^tV\Sigma_1U^tb \\ &= V\Sigma^t\Sigma\Sigma_1U^tb \\ &= V\Sigma^tU^tb \\ &= A^+b. \end{aligned}$$

(The fact that $\Sigma^t\Sigma\Sigma_1 = \Sigma^t$ is easy to check.)

This shows that $x = A^+b$ is a solution of the least squares problem.

The system of normal equations $A^tAx = A^tb$ is a system of n linear equations in n unknowns and thus, as seen in chapter 1, it has a unique solution if and only if A^tA is invertible. Now A^tA is invertible if and only if rank $A^tA = n$ and rank A^tA = rank A by lemma (8.3.4). Thus the least squares problem has a unique solution if and only if rank A = n.

For the rest of the proof let rank A = r where $r < n$. Using the same notation as earlier and writing $y = V^tx$ we see that

$$\| Ax - b \| = \| U\Sigma V^tx - b \| = \| \Sigma y - U^tb \|$$

since $\| \ \|$ is unaltered by the orthogonal transformation U^t.

Thus x minimizes $\| Ax - b \|$ if and only if y minimizes $\| \Sigma y - c \|$ where $y = V^tx$ and $c = U^tb$.

The matrix Σ has the non-zero singular values $\sigma_1, \sigma_2, \ldots, \sigma_r$ of A in the places (1,1),(2,2), . . . ,(r,r) and zero elsewhere. Writing c and y as column vectors (c_i) and (y_i) we obtain

$$\| \Sigma y - c \|^2 = \sum_{i=1}^{r} (\sigma_i y_i - c_i)^2 + \sum_{i=r+1}^{n} c_i^2$$

As y varies it is clear that the minimum value of $\| \Sigma y - c \|^2$ will be $\sum_{i=r+1}^{n} c_i^2$ which occurs when $y_i = c_i/\sigma_i$ for $i = 1,2,\ldots,r$, the rest of the y_i being free to take any value. Since $r < n$ we have an infinite set of solutions y and the value of y with minimal euclidean norm is obtained by putting $y_i = 0$ for each $i = r+1, r+2,\ldots,n$. This choice of y can be written as $y = \Sigma_1 c$.

This value of y also yields the value of x of minimal euclidean norm since $x = Vy$ and V is orthogonal. The minimal norm solution is given by $x = Vy = V\Sigma_1 c = V\Sigma_1 U^t b = A^+ b$. This completes the proof.

Problems 8E

1. Find the unique solution of the least squares problem of (8.5.1) for $A = \begin{pmatrix} 2 & -2 \\ 1 & 1 \\ 2 & 2 \end{pmatrix}$, $b = \begin{pmatrix} 500 \\ 600 \\ 900 \end{pmatrix}$.

2. Find the solution of minimal euclidean norm to the least squares problem for $A = \begin{pmatrix} 2 & 3 & 4 \\ 2 & 2 & 3 \\ 0 & 1 & 1 \end{pmatrix}$, $b = \begin{pmatrix} 100 \\ 200 \\ 300 \end{pmatrix}$.

3. Let $A = (a_{ij})$ be an m×n matrix , $x = (x_i)$ a column vector in $\mathbb{R}^n$, and $b = (b_i)$ a column vector in $\mathbb{R}^m$.

Show that $\| Ax - b \|^2 = \sum_{k=1}^{m} (\sum_{j=1}^{n} a_{kj} x_j - b_k)^2$.

Determine the partial derivatives $\partial f/\partial x_i$, $i = 1,2,\ldots,n$, of the function $f(x_1,x_2,\ldots,x_n) = \| Ax - b \|^2$.

Show that the equations $\partial f/\partial x_i = 0$ for $i = 1,2,\ldots,n$ are exactly the same as the normal equations $A^t A x = A^t b$.

8.6 Gauss-Seidel iteration

8.6.1 Sparse matrices

Systems of linear equations $Ax = b$ encountered in practical problems can often involve a *sparse matrix* A, i.e. a matrix A for which the great majority of the entries are equal to zero. One place where these tend to occur is in the numerical solution of partial differential equations. The method of Gaussian elimination for solving $Ax = b$ may quickly destroy the sparseness and so may not be the most efficient numerical method of solution. In order to take advantage of the sparseness, to save storage space on the computer etc., iterative methods of solution are commonly used for the numerical solution of $Ax = b$ where A is sparse. We confine ourselves to a description of one of the best-known iterative methods, namely the *Gauss-Seidel method.*

8.6.2 The Gauss-Seidel method

We consider a system $Ax = b$ of n linear equations in n unknowns x_i, $i = 1,2,. ,n$. As in chapter 1 we write $A = (a_{ij})$, $x = (x_i)$, and $b = (b_i)$ which are respectively $n \times n$, $n \times 1$, and $n \times 1$ matrices with real entries. Suppose that A is invertible so that, by (1.3.6), the system $Ax = b$ has a unique solution which is given by $x = A^{-1}b$. To apply our iterative method we will further suppose that each diagonal entry a_{ii} of A is non-zero.

The general idea of an iterative method of solving $Ax = b$ is as follows;

We make an initial guess at a solution by choosing some vector in $\mathbb{R}^n$ which we denote $x^{(0)}$. (Take $x^{(0)} = 0$ if no better guess is available.) We then generate a sequence of vectors $x^{(k)}$ in $\mathbb{R}^n$, $k = 0,1,2, \ldots$, each $x^{(k)}$ being obtained from $x^{(k-1)}$ by some formula. If our iterative method is to be of any use then we would hope that, as k tends to infinity, the sequence of vectors $x^{(k)}$ converges to the true solution $x = A^{-1}b$ of the system and that this convergence is valid no matter what our initial guess $x^{(0)}$ happened to be.

Since we are assuming that each a_{ii} is non-zero it is easy to see that the system of n linear equations $Ax = b$ can be rewritten as follows;

$$x_1 = a_{11}^{-1}(b_1 - a_{12}x_2 - a_{13}x_3 - \ldots\ldots - a_{1n}x_n)$$

$$x_2 = a_{22}^{-1}(b_2 - a_{21}x_1 - a_{23}x_3 - \ldots\ldots - a_{2n}x_n)$$

...

...

...

...

$$x_n = a_{nn}^{-1}(b_n - a_{n1}x_1 - a_{n2}x_2 - \ldots\ldots - a_{n(n-1)}x_{n-1})$$

These n equations can be written more concisely in the form

$$x_j = a_{jj}^{-1}\,(b_j - \sum_{\substack{i=1\\ i\neq j}}^{n} a_{ji}x_i) \text{ for } j = 1,2,\ldots,n.$$

These equations are the motivation for the iterative method which we will now describe;

The *Jacobi iteration* for the solution of $Ax = b$ is the sequence of vectors $x^{(k)} \in \mathbb{R}^n$, the components of $x^{(k)}$ being denoted $x_j^{(k)}$, $j = 1,2,\ldots,n$ defined recursively by the equations

$$x_j^{(k+1)} = a_{jj}^{-1}\,(b_j - \sum_{\substack{i=1\\ i\neq j}}^{n} a_{ji}x_i^{(k)})$$

for $j = 1,2,\ldots,n$. To obtain the components of $x^{(k+1)}$ we put $x_i = x_i^{(k)}$ for each $i = 1,2,\ldots,n$ into the right-hand-side of the set of n equations given earlier.

The *Gauss-Seidel* iteration for the solution of $Ax = b$ is the sequence of vectors $x^{(k)} \in \mathbb{R}^n$, the components of $x^{(k)}$ being denoted $x_j^{(k)}$ for $j = 1,2,\ldots,n$, defined recursively by the equations

$$x_j^{(k+1)} = a_{jj}^{-1} \left(b_j - \sum_{i=1}^{j-1} a_{ji} x_i^{(k+1)} - \sum_{i=j+1}^{n} a_{ji} x_i^{(k)}\right)$$

for $j = 1,2,\ldots,n$.

The Gauss-Seidel iteration updates the Jacobi iteration in the sense that in the calculation of $x_j^{(k+1)}$ it uses the values already calculated for $x_i^{(k+1)}$, $i = 1,2,\ldots,j-1$.

A succinct matrix representation of the formula for the Gauss-Seidel iteration can be obtained as follows;

Write $A = L + U$ where L has the same entries as A on and below the diagonal but zero elsewhere while U has the same entries as A above the diagonal but zero on and below the diagonal. Thus L is lower triangular, and U is upper triangular with all diagonal entries zero. Then the Gauss-Seidel iteration is given by

$$x^{(k+1)} = L^{-1} (b - Ux^{(k)}).$$

Note that L is invertible because the diagonal entries of A are assumed to be non-zero. The reader should verify this formula by showing that

$$Lx^{(k+1)} = b - Ux^{(k)}$$

It is vitally important to know exactly when the Gauss-Seidel iteration converges to the unique solution of the system. The eigenvalues of the matrix $M = -L^{-1}U$ play a key role in this.

8.6.3 **Theorem**

Let A be an invertible $n{\times}n$ matrix with real entries and suppose that each diagonal entry a_{ii} is non-zero. Write $A = L + U$ as above and let $M = -L^{-1}U$. Then the Gauss-Seidel iteration $x^{(k)}$ for the system $Ax = b$

converges to the solution $x = A^{-1}b$ for every initial guess $x^{(0)}$ if and only if $|\lambda| < 1$ for each eigenvalue λ of M.

Proof

Write $d^{(k)} = x^{(k)} - x$, i.e. $d^{(k)}$ is the difference between the k-th iterate and the true solution. The Gauss-Seidel iteration converges to x if and only if $d^{(k)}$ tends to the zero vector as k tends to infinity.

From the formula for the Gauss-Seidel iteration we have that $Lx^{(k+1)} = b - Ux^{(k)}$. Thus $Ld^{(k+1)} = b - Ux^{(k)} - Lx$

$$= b - U(x^{(k)} - x) - Ux - Lx$$

$$= -Ud^{(k)}$$

This shows that $d^{(k+1)} = Md^{(k)}$ for each $k = 0,1,2, \ldots$, so that $d^{(k)} = M^k d^{(0)}$ for each k.

As k tends to infinity if M^k tends to the zero matrix then clearly $d^{(k)}$ tends to the zero vector. Conversely if $d^{(k)}$ tends to the zero vector as k tends to infinity for every initial guess $x^{(0)}$, (i.e. for every vector $d^{(0)}$ in $\mathbb{R}^n$), then M^k must tend to the zero matrix as k tends to infinity.

We saw in problem 1 of Problems 5B that, by examining the Jordan form, M^k tends to the zero matrix as k tends to infinity if and only if $|\lambda| < 1$ for each eigenvalue λ of M. This completes the proof.

8.6.4 **Exercise**

Show that the eigenvalues of the matrix $M = -L^{-1}U$ above are precisely the solutions of the equation $\det(\lambda L + U) = 0$, i.e the solutions of

$$\det \begin{pmatrix} \lambda a_{11} & a_{12} & a_{13} & & a_{1n} \\ \lambda a_{21} & \lambda a_{22} & a_{23} & & a_{2n} \\ & & & & \\ \lambda a_{n1} & \lambda a_{n2} & & & \lambda a_{nn} \end{pmatrix} = 0.$$

8.6.5 **Theorem**

Let A be a positive-definite real symmetric $n{\times}n$ matrix. The Gauss-Seidel iteration always converges to the solution $x = A^{-1}b$ for any system of n linear equations $Ax = b$ in n unknowns.

Proof

Since A is positive-definite A is invertible and each diagonal entry a_{ii} is positive. By (8.6.3) it suffices to show that each eigenvalue λ of $M = -L^{-1}U$ satisfies $|\lambda| < 1$ where we write $A = L + U$ as in (8.6.3).

$M = -L^{-1}U = -L^{-1}(A - L) = I - L^{-1}A$. Let λ be an eigenvalue of M with a corresponding eigenvector v. Then $(I - L^{-1}A)v = \lambda v$. This implies that $(1 - \lambda)v = L^{-1}Av$, i.e. $(1 - \lambda)Lv = Av$. Hence $v^tAv = (1 - \lambda)v^tLv$.

Taking the conjugate transpose of this last equation and noting that A and L are real and that A is symmetric yields that $v^tAv = (1 - \bar{\lambda})v^tL^tv$.

Let D be the diagonal matrix with entries a_{ii} on the diagonal and zero elsewhere. Note that D is positive-definite and that $D = L + L^t - A$. Then

$$\begin{aligned} v^tDv &= v^tLv + v^tL^tv - v^tAv \\ &= [(1 - \lambda)^{-1} + (1 - \bar{\lambda})^{-1} - 1]v^tAv. \end{aligned}$$

(Note that $1 - \lambda = 0$ is impossible as it would imply $v^tAv = 0$ contradicting the positive-definiteness of A, v being a non-zero vector. Similarly $1 - \bar{\lambda} = 0$ is impossible.)

Since A and D are each positive-definite this ensures that

$$(1 - \lambda)^{-1} + (1 - \bar{\lambda})^{-1} - 1 > 0.$$

Multiplying this last inequality by $(1 - \lambda)(1 - \bar{\lambda})$ we find that

$$1 - \lambda + 1 - \bar{\lambda} - (1 - \lambda)(1 - \bar{\lambda}) > 0$$

Hence $1 - \lambda\bar{\lambda} > 0$ so that $|\lambda| < 1$ and the proof is complete.

8.6.6 **The rate of convergence of the Gauss-Seidel iteration**

Recall that the *spectral radius* of a square matrix M is the non-negative real number $\rho(M) = \max\{|\lambda|: \lambda \text{ is an eigenvalue of } M\}$ (See problem 3 of Problems 7B.)

Theorem (8.6.3) says that the Gauss-Seidel iteration converges if and only if $\rho(M) < 1$. In the proof of (8.6.3) we saw that $d^{(k)} = M^k d^{(0)}$ and so the rate at which $x^{(k)}$ converges to the solution x will depend on how quickly the matrix power M^k tends to the zero matrix. The rate of convergence of M^k to zero is determined by $\rho(M)$. If $\rho(M)$ is close to 1 the convergence will be slow while if $\rho(M)$ is close to zero the convergence will be rapid. (This can be seen by considering the Jordan form J of M. The Jordan form J has the eigenvalues of M on the diagonal and the convergence of J^k to zero is determined by the convergence of the powers of these eigenvalues.) Note also that if $\rho(M) = 0$, which implies that M is nilpotent, then $d^{(k)}$ becomes zero for a finite value of k so that the true solution is obtained after finitely many steps of the Gauss-Seidel iteration.

8.6.7 **Example**

Let $A = (a_{ij})$ be the 100x100 matrix with entries given by

$$a_{ii} = 1 \text{ for all } i,$$

$$a_{i(i+1)} = 1 \text{ for all } i,$$

$$a_{ij} = 0 \text{ otherwise.}$$

Let $b = (b_i) \in \mathbb{R}^{100}$ be given by $b_i = 2$ for all i.

It is not difficult to see directly that the system $Ax = b$ has a unique solution $v \in \mathbb{R}^n$, $v = (v_i)$ where $v_i = 2$ for all i even, and $v_i = 0$ for all i odd.

The reader may check that the Gauss-Seidel iteration, with initial guess $x^{(0)} = 0$, yields that $x^{(100)} = v$.

It turns out that $x_{100}^{(2)} = 2$, $x_i^{(2)} = 0$ otherwise,

$$x_{100}^{(4)} = x_{98}^{(4)} = 2,\ x_i^{(4)} = 0 \text{ otherwise},$$

$$x_{100}^{(6)} = x_{98}^{(6)} = x_{96}^{(6)} = 2,\ x_i^{(6)} = 0 \text{ otherwise, etc.}$$

(The matrix $M = -L^{-1}U$ for this example is nilpotent of index 100.)

8.6.8 **Successive Over-Relaxation (The SOR method)**

We have seen that when the spectral radius of the matrix M in the Gauss-Seidel iteration is close to 1 the rate of convergence is slow. The following modification of the Gauss-Seidel method, which is known as the method of *successive over-relaxation*, can significantly improve the rate of convergence. For short it is often called the *SOR-method.*

Let $A = (a_{ij})$ be as in the Gauss-Seidel method. Let ω be a non-zero real number. Let L_ω be the lower triangular n×n matrix with entries a_{ii} on the diagonal and ωa_{ij} in each (i,j)-place below the diagonal. Let U_ω be the upper triangular n×n matrix with entries $(\omega - 1)a_{ii}$ on the diagonal and ωa_{ij} in each (i,j)-place above the diagonal.

Define an iteration $(x^{(k)})$ for the system $Ax = b$ by requiring

$$x^{(k+1)} = L_\omega^{-1}(\omega b - U_\omega x^{(k)})$$

Note that $L_\omega + U_\omega = \omega A$ and that the system $Ax = b$ has precisely the same solution as the system $\omega Ax = \omega b$ since $\omega \neq 0$. The iterative process above will, in the same way as in (8.6.3), converge to the solution of $Ax = b$ if and only if the matrix $M_\omega = -L_\omega^{-1}U_\omega$ has spectral radius less than 1. (Observe that when $\omega = 1$ the above iteration is the usual Gauss-Seidel iteration.)

By choosing ω so as to minimize the spectral radius $\rho(M_\omega)$ we should in general obtain a faster rate of convergence than in the ordinary Gauss-Seidel iteration. See problem 3 of Problems 8F for an example of this.

Problems 8F

1. Let A be the tridiagonal matrix $\begin{pmatrix} 2 & 1 & 0 \\ 1 & 2 & 1 \\ 0 & 1 & 2 \end{pmatrix}$ and let $b = \begin{pmatrix} 1 \\ 2 \\ 3 \end{pmatrix}$. Verify that $\begin{pmatrix} 1/2 \\ 0 \\ 3/2 \end{pmatrix}$ is the solution of $Ax = b$. Using the Gauss-Seidel method for $Ax = b$ with initial guess $x^{(0)} = \begin{pmatrix} 0 \\ 0 \\ 0 \end{pmatrix}$ determine $x^{(5)}$ and show that the error $d^{(5)}$ after 5 steps satisfies $\| d^{(5)} \| = 3/64$ where $\| \ \|$ is the cartesian norm.
Deduce that $\| d^{(k)} \| = 3/2^{k+1}$ for each k.

2. If the n×n matrix A is *strictly diagonally dominant* in the sense that $|a_{ii}| > \sum_{\substack{j=1 \\ j \neq i}}^{n} |a_{ij}|$ for each $i = 1,2,\ldots,n$, show that the Gauss-Seidel iteration for $Ax = b$ will always converge.
[*Hint* - let λ be an eigenvalue of $M = -L^{-1}U$ and let $v = (v_i)$ be an eigenvector for λ. Choose v so that $\max |v_i| = 1$, i.e. v has cartesian norm one. Deduce from the equations $Uv + \lambda Lv = 0$ that $|\lambda||v_i| < 1$ for each $i = 1,2,\ldots,n$ and hence that $|\lambda| < 1$.]

3. Let $A = \begin{pmatrix} 11 & 1 \\ 10 & 1 \end{pmatrix}$. Using our earlier notation from (8.6.2) and (8.6.8) show that $\rho(M) = 0.909$ for $M = -L^{-1}U$ where $A = L + U$, that $\rho(M_\omega) = 0.537$ for $\omega = (11 - \sqrt{11})/5$, and that this is the minimum value of $\rho(M_\omega)$ as ω varies through all non-zero real numbers.
[*Hint* - the characteristic polynomial of M_ω is a quadratic and the minimal value of $\rho(M_\omega)$ will occur when this quadratic has equal real roots.]

REFERENCES AND FURTHER READING

[C] S.L.Campbell, Recent applications of generalized inverses, Pitman, London, 1982.

[GVL] G.H.Golub and C.F.Van Loan, Matrix computations (2nd. edition), Johns Hopkins University Press, Baltimore-London, 1989.

[HJ] R.A.Horn and C.R.Johnson, Matrix analysis, Cambridge University Press, Cambridge, 1985.

[N] M.Nashed, Generalized inverses and applications, Academic Press, New York, 1976.

[R] W.Rudin, Real and complex analysis (2nd. edition), McGraw-Hill, New York, 1974.

[Se] E.Seneta, Non-negative matrices and Markov chains (2nd. edition), Springer, Berlin, 1981.

[St] G.Strang, Linear algebra and its applications (2nd. edition), Academic Press, London-New York, 1980.

ANSWERS TO SELECTED PROBLEMS

Problems 1A

1. $AB = \begin{pmatrix} 20 & -2 \\ 12 & -1 \end{pmatrix}$, $DC = \begin{pmatrix} 2 & 7 \\ -16 & 14 \\ -7 & -2 \end{pmatrix}$, $A^2 = \begin{pmatrix} 7 & 12 \\ 4 & 7 \end{pmatrix}$, $A^3B^2 = \begin{pmatrix} 1032 & -284 \\ 596 & -164 \end{pmatrix}$.

Problems 1B

1. (i) (-1/2,-117/2,33,4), (ii) $\{(11\alpha/3 + 10/3, -11\alpha - 10, 6\alpha + 6, \alpha); \alpha \in \mathbb{R}\}$

2. (1,0,0,0)

3. (a) Inconsistent, (b) $\{(\alpha, (1 - \alpha)/3, (2 - 2\alpha)/3); \alpha \in \mathbb{R}\}$

5. (a) (3,1,1), (b) $\{(\alpha, 0, \alpha - 2); \alpha \in \mathbb{R}\}$

Problems 1C

1. $\begin{pmatrix} 10 & 7 & -3 \\ -3 & -2 & 1 \\ 17 & 11 & -5 \end{pmatrix}$, $\begin{pmatrix} 1 & -2 & 1 & 0 \\ 0 & 1 & -2 & 1 \\ 0 & 0 & 1 & -2 \\ 0 & 0 & 0 & 1 \end{pmatrix}$, $\begin{pmatrix} 1/2 & -1/2 & 0 \\ 0 & 1/2 & -1/2 \\ 0 & 0 & 1/2 \end{pmatrix}$

4. 6,0,-40

6. $(a + b + c + d)(a + ib - c - id)(a - b + c - d)(a - ib - c + id)$

Problems 1E

2. $k = 1$, $\{(2, \alpha, -1-\alpha); \alpha \in \mathbb{R}\}$

4. $\begin{pmatrix} 1 & 0 & 0 \\ -1 & 1 & 0 \\ 4 & -1 & 1 \end{pmatrix}\begin{pmatrix} 1 & 2 & 4 \\ 0 & -1 & 7 \\ 0 & 0 & 5 \end{pmatrix}$

Problems 2A

1. (i) Yes, (ii) No, (iii) Yes, (iv) No.

Problems 2B

1. (i) Yes, (ii) Yes, (iii) Yes.

6. (i) Yes, (ii) No, (iii) Yes.

Problems 2C

3. $n(n+1)/2$, $n(n-1)/2$

6. 2,3,1

7. 2,3,1

Problems 2D

2. $\begin{pmatrix}1 & 1 & 0\\ 2 & -1 & -1\\ 1 & 1 & 1\end{pmatrix}$, $M_{B_1B_2}(1) = \begin{pmatrix}-1 & 1 & 1\\ 1 & 1 & 1\\ 1 & 0 & 1\end{pmatrix}$, $M_{B_2B_1}(1) = \begin{pmatrix}-1/2 & 1/2 & 0\\ 0 & 1 & -1\\ 1/2 & -1/2 & 1\end{pmatrix}$

$M_{B_1B_1}(f) = \begin{pmatrix}-1 & 1 & 2\\ 1 & -1 & -3\\ 1 & 2 & 3\end{pmatrix}$, $M_{B_2B_2}(f) = \begin{pmatrix}-9/2 & 0 & -5/2\\ -9 & -3 & -9\\ 17/2 & 3 & 17/2\end{pmatrix}$

4. rank 3, nullity 1, dim $I \cap U = 1$

5. $\begin{pmatrix}1 & 3 & 5 & 7\\ 0 & 2 & 6 & -5\\ 0 & -1 & 0 & 2\\ 0 & -1 & -1 & 5\end{pmatrix}$

6. rank 6, nullity 1

Problems 2E

12. (i) $\{1/\sqrt{3}(1,1,1), 1/\sqrt{6}(1,-2,1), 1/\sqrt{2}(1,0,-1)\}$

(ii) $\{1/\sqrt{6}(1,1,1), 1/\sqrt{12}(1,-2,1), 1/\sqrt{12}(3,0,-1)\}$

(iii) $\{1/\sqrt{2}, \sqrt{(3/2)}x, 3x^2/4 - 1/4, \sqrt{(7/2)}(5x^3/2 - 3x/2)\}$

Problems 4A

1. (i) $\lambda = 4,2,-2$, each of alg. mult. one

(ii) $\lambda = 1$, alg. mult. two and geom. mult. two, $\lambda = -1$, alg. mult. one

(iii) $\lambda = 1, 1+2\sqrt{2}, 1-2\sqrt{2}$, each of alg. mult. one

(iv) $\lambda = 1, 1+i$, each of alg. mult. one

(v) $\lambda = 1+i$, alg. mult two and geom. mult. two, $\lambda = 1$, alg. mult. one

(vi) $\lambda = 0$, alg. mult. two and geom. mult. two, $\lambda = 2$, alg. mult. two and geom. mult. two

5. $\lambda = n$, alg. mult one, $\lambda = 0$, alg. mult. n-1

Problems 4C

3. (i) $(\lambda-2)^2$, (ii) λ^2, (iii) $\lambda(\lambda-2)\lambda+2)$, (iv) $(\lambda-2)^2(\lambda-5)$, (v) $\lambda(\lambda-3)$

Problems 4D

3. $A^6 = \begin{pmatrix}379 & -315 & 625\\ 378 & -314 & 625\\ 0 & 0 & 1\end{pmatrix}$

Problems 5A

2. $\lambda = 6$, one 2x2 block and two 1x1 blocks, $\lambda = 7$, either two 2x2 blocks or one 2x2 block and two 1x1 blocks.

6. $\lambda = 2$, one 4x4 block and one 1x1 block, $\lambda = -1$, one 1x1 block.

7. $\lambda = 1$, one 2x2 block, $\lambda = -2$, one 1x1 block, $\lambda = -3$, one 1x1 block.

Problems 5B

2. $\lambda = 1$, one 3x3 block and two 2x2 blocks.

Problems 5C

1. $x_1 = 10e^{3t} - 6e^{-t}$, $x_2 = 5e^{3t} + 3e^{-t}$

3. $x = 3A\cos t + 3B\sin t + 2Ce^{\sqrt{3}t} + 2De^{-\sqrt{3}t}$,

   $y = A\cos t + B\sin t + Ce^{\sqrt{3}t} + De^{-\sqrt{3}t}$.

5. 35.06%

7. 62.8%, 62.5%

Problems 6C

3. (i) No, (ii) Yes, (iii) No.

Problems 6D

1. (i) 1, (ii) -1, (iii) 1.

3. Congruent, each has signature zero.

Problems 6E

2. $\lambda_1 = 1$, $\lambda_2 = (1 + \sqrt{13})/2$, $\lambda_3 = (1 - \sqrt{13})/2$.

4. $A = \begin{pmatrix} 8 & -3 \\ -3 & 3 \end{pmatrix}$, $B = \begin{pmatrix} 0 & 6 \\ 6 & 6 \end{pmatrix}$, solution $x = Py$ where $P = \begin{pmatrix} 1/2\sqrt{2} & 3/2\sqrt{30} \\ 1/\sqrt{2} & 1/\sqrt{30} \end{pmatrix}$ and

$$y = \begin{pmatrix} K_1\cos\sqrt{6}t + K_2\sin\sqrt{6}t \\ K_3e^{\sqrt{2/5}t} + K_4e^{-\sqrt{2/5}t} \end{pmatrix}.$$

Problems 7C

1. (i) 206.766, (ii) 162.5

3. $c(A) = (\alpha+1)/\alpha$ for $0 < \alpha \leq 1/2$, $c(A) = |(\alpha+1)/(\alpha-1)|$ for $\alpha \geq 1/2$.

Problems 8C

1.(i) $\Sigma = \begin{pmatrix} \sqrt{10} & 0 \\ 0 & \sqrt{8} \\ 0 & 0 \end{pmatrix}$, (ii) $\Sigma = \begin{pmatrix} \sqrt{12} & 0 & 0 \\ 0 & 0 & 0 \end{pmatrix}$, (iii) $\Sigma = \begin{pmatrix} \sqrt{45} & 0 & 0 \\ 0 & \sqrt{18} & 0 \\ 0 & 0 & \sqrt{5} \end{pmatrix}$.

Problems 8D

1. (i) $\begin{pmatrix} 1/4 & 1/10 & 1/5 \\ -1/4 & 1/10 & 1/5 \end{pmatrix}$, (ii) $\begin{pmatrix} 1/6 & -1/6 \\ 1/12 & -1/12 \\ -1/12 & 1/12 \end{pmatrix}$, (iii) $\begin{pmatrix} 1/18 & -2/5 & 2/45 \\ 1/9 & 0 & -1/9 \\ 1/9 & 1/5 & 4/45 \\ 1/6 & 0 & 0 \end{pmatrix}$

Problems 8E

1. $\begin{pmatrix} 365 \\ 115 \end{pmatrix}$

2. $\begin{pmatrix} -525/2 \\ 225 \\ 25 \end{pmatrix}$

INDEX